# Communications
# in Computer and Information Science    1623

T0235638

More information about this series at https://link.springer.com/bookseries/7899

Greg H. Parlier · Federico Liberatore ·
Marc Demange (Eds.)

# Operations Research and Enterprise Systems

9th International Conference, ICORES 2020, Valetta, Malta,
February 22–24, 2020, and 10th International Conference
ICORES 2021, Virtual Event, February 4–6, 2021
Revised Selected Papers

 Springer

*Editors*
Greg H. Parlier
INFORMS
Catonsville, MD, USA

Federico Liberatore
UC3M-BS Institute of Financial Big Data
Universidad Carlos III de Madrid
Getafe, Madrid, Spain

Marc Demange
School of Science-Mathematical Sciences
RMIT University
Melbourne, VIC, Australia

ISSN 1865-0929          ISSN 1865-0937 (electronic)
Communications in Computer and Information Science
ISBN 978-3-031-10724-5          ISBN 978-3-031-10725-2 (eBook)
https://doi.org/10.1007/978-3-031-10725-2

This Springer imprint is published by the registered company Springer Nature Switzerland AG
The registered company address is: Gewerbestrasse 11, 6330 Cham, Switzerland

# Preface

This book includes extended and revised versions of selected papers from the 9th and 10th editions of the International Conference on Operations Research and Enterprise Systems (ICORES 2020 and ICORES 2021). ICORES 2020 was held in Valletta, Malta during February 22–24, 2020, and ICORES 2021 was held as an online event, due to the COVID-19 pandemic, during February 4–6, 2021 2021.

ICORES 2020 received 68 paper submissions of which 7% are included in this book, and ICORES 2021 received 64 paper submissions of which 9% are also included in this book. These papers were selected based on several criteria including reviews provided by Program Committee members, session chair assessments, and also program chair perspectives across all papers included in the technical program. The authors of these selected papers were then invited to submit revised and extended versions of their papers for formal publication.

The purpose of the annual International Conference on Operations Research and Enterprise Systems (ICORES) is to bring together researchers, engineers, and practitioners interested in both advances and applications in the field of operations research. Two simultaneous tracks are held, one covering domain independent methodologies and technologies and the other practical work developed in specific application areas.

The papers selected for this book contribute to operations research and our better understanding of complex enterprise systems. We commend each of the authors for their contributions, and gratefully thank our many reviewers who ensured the high quality of this publication.

February 2021

Greg H. Parlier
Federico Liberatore
Marc Demange

# Organization

## Conference Chair

Marc Demange　　　　　　　　RMIT University, Australia

## Program Co-chairs

Greg H. Parlier　　　　　　　North Carolina State University, USA
Federico Liberatore　　　　　Cardiff University, UK

## Program Committee

**Served in 2020**

Alberto Tonda　　　　　　　INRAE, France
Ana Rocha　　　　　　　　　Universidade do Minho, Portugal
Andrea Scozzari　　　　　　Università degli Studi "Niccolò Cusano", Italy
Angelo Sifaleras　　　　　　University of Macedonia, Greece
Arij Lahmar　　　　　　　　University of Dubai, UAE
Bisera Andric Gušavac　　　University of Belgrade, Serbia
Cem Saydam　　　　　　　　University of North Carolina at Charlotte, USA
Claudio Gentile　　　　　　Institute for System Analysis and Computer
　　　　　　　　　　　　　　Science, CNR-IASI, Italy
Daniel Karapetyan　　　　　University of Essex, UK
Dario Landa-Silva　　　　　University of Nottingham, UK
David Bergman　　　　　　　University of Connecticut, USA
Dries Goossens　　　　　　　Ghent University, Belgium
Eduardo Barbosa　　　　　　Brazilian National Institute for Space Research
　　　　　　　　　　　　　　(INPE), Brazil
Florbela Correia　　　　　　Instituto Politécnico de Viana do Castelo, Portugal
Gaia Nicosia　　　　　　　　Università degli Studi Roma Tre, Italy
Giuseppe Lancia　　　　　　University of Udine, Italy
Giuseppe Stecca　　　　　　Institute of Systems Analysis and Computer
　　　　　　　　　　　　　　Science, CNR-IASI, Italy
Godichaud Matthieu　　　　Université de Technologie de Troyes, France
Guzmán Santafé Rodrigo　　Universidad Pública de Navarra, Spain
Il-Gyo Chong　　　　　　　　Samsung Electronics, South Korea
Janny Leung　　　　　　　　The Chinese University of Hong Kong, Shenzhen,
　　　　　　　　　　　　　　China
Jin-Kao Hao　　　　　　　　University of Angers, France

| Kurt Engemann | Iona College, USA |
| Laurent Houssin | LAAS-CNRS, France |
| Lotte Berghman | Toulouse Business School, France |
| Maria Eugénia Captivo | Universidade de Lisboa, Portugal |
| Maria Isabel Gomes | Universidade NOVA de Lisboa, Portugal |
| Marin Lujak | IMT Nord Europe, France |
| Marta Gomes | Instituto Superior Técnico, Universidade de Lisboa, Portugal |
| Marta Mesquita | Universidade de Lisboa, Portugal |
| Mohamed Saleh | Cairo University, Egypt |
| Nesim Erkip | Bilkent University, Turkey |
| Nikolai Dokuchaev | Curtin University, Australia |
| Paolo Ventura | Consiglio Nazionale delle Ricerche, Italy |
| Patrick Siarry | LiSSi, Université Paris Est Créteil, France |
| Pedro Oliveira | Universidade do Porto, Portugal |
| Prabhat Mahanti | University of New Brunswick, Canada |
| Renato Bruni | Sapienza University of Rome, Italy |
| Renaud De Landtsheer | CETIC, Belgium |
| Selin Özpeynirci | Izmir University of Economics, Turkey |
| Stefano Smriglio | University of L'Aquila, Italy |
| Stella Kapodistria | Eindhoven University of Technology, The Netherlands |
| Ulrich Pferschy | University of Graz, Austria |
| Veronica Piccialli | Tor Vergata University of Rome, Italy |
| Viliam Makis | University of Toronto, Canada |
| Xavier Gandibleux | University of Nantes, France |
| Yan Pengyu | University of Electronic Science and Technology of China, China |
| Yi Qu | University of Northumbria, UK |

## Served in 2021

| Günther Raidl | TU Wien, Austria |
| Roberto Montemanni | University of Modena and Reggio Emilia, Italy |
| Shaligram Pokharel | Qatar University, Qatar |

## Served in 2020 and 2021

| Ahmed Bufardi | Independent Researcher, Switzerland |
| Ahmed Kheiri | Lancaster University, UK |
| Alfonso Mateos Caballero | Universidad Politécnica de Madrid, Spain |
| Ali Emrouznejad | Aston University, UK |
| Andre Cire | University of Toronto, Canada |
| Andre Rossi | Université Paris-Dauphine, France |

| | |
|---|---|
| Andrea D'Ariano | Roma Tre University, Italy |
| Antonio Jiménez-Martín | Universidad Politécnica de Madrid, Spain |
| Bernardetta Addis | Université de Lorraine, France |
| Bo Chen | University of Warwick, UK |
| Boris Goldengorin | Moscow University of Physics and Technology, Russia |
| Cameron Walker | University of Auckland, New Zealand |
| Carlo Filippi | University of Brescia, Italy |
| Carlo Meloni | Sapienza Università di Roma, Italy |
| Celso Ribeiro | Universidade Federal Fluminense, Brazil |
| Chefi Triki | Hamad Bin Khalifa University, Qatar |
| Chenyi Hu | University of Central Arkansas, USA |
| Christelle Guéret | University of Angers, France |
| Clarisse Dhaenens | University of Lille, France |
| Concepción Maroto | Universidad Politécnica de Valencia, Spain |
| Cristiana Silva | University of Aveiro, Portugal |
| Cyril Briand | LAAS-CNRS, France |
| Edson Luiz França Senne | Universidade Estadual Paulista, Brazil |
| Elena Tànfani | Università degli Studi di Genova, Italy |
| El-Houssaine Aghezzaf | Ghent University, Belgium |
| Emanuele Manni | University of Salento, Italy |
| Fabrizio Marinelli | Università Politecnica delle Marche, Italy |
| Fabrizio Rossi | University of L'Aquila, Italy |
| Francesca Guerriero | University of Calabria, Italy |
| Gerhard Woeginger | RWTH Aachen, Germany |
| Giancarlo Bigi | University of Pisa, Italy |
| Gianpiero Bianchi | Italian National Institute of Statistics, Italy |
| Giorgio Gnecco | IMT School for Advanced Studies Lucca, Italy |
| Giovanni Fasano | Ca' Foscari University of Venice, Italy |
| Han Hoogeveen | Universiteit Utrecht, The Netherlands |
| Hanno Hildmann | TNO, The Netherlands |
| Heliodoro Cruz-Suárez | Universidad Juárez Autónoma de Tabasco, Mexico |
| Inneke Van Nieuwenhuyse | KU Leuven, Belgium |
| J. M. van den Akker | Utrecht University, The Netherlands |
| Jairo Montoya-Torres | Universidad de La Sabana, Colombia |
| Javier Alcaraz | Universidad Miguel Hernandez de Elche, Spain |
| Johann Hurink | University of Twente, The Netherlands |
| José Oliveira | Universidade do Minho, Portugal |
| Josef Jablonsky | Prague University of Economics and Business, Czech Republic |
| Kaushik Das Sharma | University of Calcutta, India |

| | |
|---|---|
| Kenji Hatano | Doshisha University, Japan |
| Laura Scrimali | University of Catania, Italy |
| Lionel Amodeo | University of Technology of Troyes, France |
| Luis Miguel Ferreira | Universidade de Coimbra, Portugal |
| Maria Vlasiou | Eindhoven University of Technology, The Netherlands |
| Mauro Passacantando | University of Pisa, Italy |
| Michal Koháni | University of Zilina, Slovakia |
| Michela Robba | University of Genova, Italy |
| Mirela Danubianu | "Stefan cel Mare" University of Suceava, Romania |
| Mirko Cesarini | University of Milano-Bicocca, Italy |
| Monica Gentili | University of Louisville, USA |
| Muhammad Marwan Muhammad Fuad | Coventry University, UK |
| Paola Festa | University of Naples Federico II, Italy |
| Patrizia Daniele | University of Catania, Italy |
| Pedro Martins | Polytechnic Institute of Coimbra, Portugal |
| Pierre L'Ecuyer | Université de Montréal, Canada |
| Pierre Lopez | LAAS-CNRS, Université de Toulouse, France |
| Renata Mansini | University of Brescia, Italy |
| Roberto Cordone | University of Milan, Italy |
| Rym M'Hallah | Kuwait University, Kuwait |
| Santoso Wibowo | Central Queensland University, Australia |
| Sara Mattia | CNR-IASI, Italy |
| Sin C. Ho | The Chinese University of Hong Kong, Hong Kong |
| Sotiria Lampoudi | Hefring Engineering, USA |
| Stefano Giordani | Tor Vergata University of Rome, Italy |
| Stefano Sebastio | United Technologies Research Centre, Ireland |
| Sujin Bureerat | Khon Kaen University, Thailand |
| Tatiana Tambouratzis | University of Piraeus, Greece |
| Tatjana Davidovic | Mathematical Institute of the Serbian Academy of Sciences and Arts, Serbia |
| Thomas Stützle | Université Libre de Bruxelles, Belgium |
| Tri-Dung Nguyen | University of Southampton, UK |
| Valentina Cacchiani | University of Bologna, Italy |
| Veljko Jeremic | University of Belgrade, Serbia |
| Wai Yuen Szeto | University of Hong Kong, Hong Kong |
| Wasakorn Laesanklang | Mahidol University, Thailand |
| Yiqiang Zhao | Carleton University, Canada |
| Yong-Hong Kuo | University of Hong Kong, Hong Kong |

# Additional Reviewers

### Served in 2020

| | |
|---|---|
| Tiziano Bacci | IASI-CNR, Italy |
| Timothy Curtois | Staff Roster Solutions, UK |
| Collin Drent | Eindhoven University of Technology, The Netherlands |
| Izuru Kume | NAIST, Japan |
| Roberto Rosetti | Università Politecnica delle Marche, Italy |
| Farshad Shams | CNRS, CentraleSupélec, France |
| Zorica Stanimirovic | University of Belgrade, Serbia |

### Served in 2021

| | |
|---|---|
| Zabih Ghelichi | University of Louisville, USA |
| Ruichao Jiang | UBC Okanagan, Canada |

### Served in 2020 and 2021

| | |
|---|---|
| Ornella Pisacane | Università Politecnica delle Marche, Italy |
| Andrea Pizzuti | Università Politecnica delle Marche, Italy |

# Invited Speakers

### 2020

| | |
|---|---|
| Tinaz Ekim | Bogaziçi University, Turkey |
| Bertrand Jouve | CNRS, France |
| Suresh P. Sethi | University of Texas at Dallas, USA |

### 2021

| | |
|---|---|
| Elena Rovenskaya | International Institute for Applied Systems Analysis, Austria |
| David Saranchak | Concurrent Technologies Corporation, USA |
| Roman Slowinski | Poznan University of Technology and Polish Academy of Sciences, Poland |

# Contents

# Approaching Set Cover Leasing, Connected Dominating Set and Related Problems with Online Deterministic Algorithms

Christine Markarian[1(✉)] and Abdul-Nasser Kassar[2]

[1] Department of Engineering and Information Technology, University of Dubai, Dubai, UAE
cmarkarian@ud.ac.ae
[2] Department of Information Technology and Operations Management,
Lebanese American University, Beirut, Lebanon
abdulnasser.kassar@lau.edu.lb

**Abstract.** Over the past decades, algorithmic study of *online* optimization problems has gained a lot of popularity in both theory and practice. The input to a classical *offline* optimization problem is given to the algorithm all at once. In an *online* optimization problem, portions of the input are revealed over time and the so-called *online algorithm* reacts to each portion, while targeting a given optimization goal against the entire input. In this paper, we study five online optimization problems that are variations of the well-known *Set Cover* (SC) and *Connected Dominating Set* (CDS) problems. SC and CDS appear in many real-world optimization scenarios emerging from social networks, wireless networks, robotics, and operations research. Two of these problems have been previously tackled with online *randomized* algorithms [2, 30]. We approach these and the other three problems with online *deterministic* algorithms, evaluated using the standard notion of *competitive analysis*. The latter measures the performance of the online algorithm against the optimal offline algorithm, that knows the entire input all at once.

**Keywords:** Online optimization problems · Connected dominating set · Set cover · Leasing · Social networks · Competitive analysis · Deterministic algorithms

## 1 Introduction

Is possible to make *provably* good optimization decisions even without knowing the future? Can the same algorithmic techniques used to solve classical optimization problems be used, if the input sequence is not entirely known in advance? These questions arise in the well-established study of *online* optimization problems, which model scenarios in which future events are impossible to know. Classically, the input to an *offline* optimization problem is given to the algorithm all upfront [3,5–7]. In an online optimization problem, portions of the input are revealed over time and the so-called *online*

---

This work is based on the conference paper [41] of ICORES 2020.

G. H. Parlier et al. (Eds.): ICORES 2020/2021, CCIS 1623, pp. 1–20, 2022.
https://doi.org/10.1007/978-3-031-10725-2_1

*algorithm* reacts to each portion, while targeting a given optimization goal against the entire input. That is, as soon as part of the input is announced, the online algorithm needs to make an optimization decision that is good enough not only against that part, but also against the entire input. This scenario seems to appear in many of our real-world scenarios, in which we are asked to perform well in the present, while having as few regrets as possible in the future. While such decisions might have minor effect in some applications, even a slight improvement in these decisions might yield to major gain in many applications, such as in stock market, rent-or-buy, or marketing scenarios.

In this paper, we design online deterministic algorithms for five online optimization problems that are online variations of two well-known problems, *Connected Dominating Set* (CDS) and *Set Cover* (SC). In CDS, we need to find a minimum connected subgraph of a given (undirected) graph such that each node is either in the subgraph or has an adjacent node in it. CDS is one of the most well-studied domination problems [22] with a wide range of applications in wireless networks [46] and social networks [4,13,21]. It is known to be $\mathcal{NP}$-complete even in planar graphs [26] and admits an $\mathcal{O}(\ln \Delta)$-approximation for general graphs, where $\Delta$ is the maximum node degree of the input graph [28]. The latter is the best possible unless $\mathcal{NP} \subseteq DTIME(n^{\log \log n})$ [23,40]. In SC, we are given a universe $\mathcal{U}$ and a collection $\mathcal{S}$ of subsets of $\mathcal{U}$, each associated with a cost, and the goal is to find subsets $\mathcal{C} \subseteq \mathcal{S}$ of minimum total cost, whose union is $\mathcal{U}$. SC is one of Karp's twenty-one problems shown to be NP-complete in 1972. It is defined in many settings and appears in various contexts [14,27,29,44]. SC has an $\mathcal{O}(\log n)$-approximation, where $n$ is the number of elements [19,24,34,39]. The latter is the best possible unless $\mathcal{P} = \mathcal{NP}$ [9].

We measure our online algorithms using the standard notion of *competitive analysis*. Given a minimization problem $X$. Let $r$ be the worst-case ratio of the cost of the solution constructed by an online algorithm $A$ to that constructed by the optimal offline algorithm that is given the entire input upfront, over all possible instances of $X$. $A$ is then *r-competitive* or has *competitive ratio r*. In other words, $A$ is at most $r$ times worse than the best solution constructed by having access to all the input, for any instance of $X$.

## 2   Overview

In this section, we give an overview of our results. The five problems addressed are the *Online Connected Dominating Set* problem (OCDS), the *Online r-hop Connected Dominating Set* problem (r-hop OCDS), the *Online Set Cover Leasing problem* (OSCL), the *Online Connected Dominating Set Leasing* problem (OCDSL), and the *Online r-hop Connected Dominating Set Leasing* problem (r-hop OCDSL). We classify these into four problem models, namely, the *online* model, the *online r-hop* model, the *online leasing* model, and the *online r-hop leasing* model.

The results for these five problems are presented in our conference paper [41]. In this work, we extend these results as follows.

- we introduce the problem models, so as many offline optimization problems may be defined in one or more of these models;

- we complete the missing proofs, which were omitted in our conference paper;
- we describe the social network applications in which these problems may arise.

Next, we define the four problem models.

## 2.1 The Online Model

The *online model* is the classical online version of an offline optimization problem. Rather than having the entire input from the beginning of the algorithm, a portion arrives in each step and the online algorithm needs to react to it. The algorithm eventually receives all or a subset of the input. When the algorithm adds a resource (subset, node, etc.) into the solution, it can be used forever, that is, throughout the entire execution of the algorithm.

While all five problems belong to the online model, the *Online Connected Dominating Set* problem (OCDS) belongs *only* to this model. We propose (in Sect. 5) the first deterministic algorithm for OCDS, with asymptotically optimal competitive ratio of $\mathcal{O}(\log^2 n)$, where $n$ is the number of nodes. The currently known result for OCDS is a randomized algorithm by Hamann *et al.* [30], with asymptotically optimal $\mathcal{O}(\log^2 n)$-competitive ratio.

## 2.2 The Online r-hop Model

The *online r-hop model* is a combination of two models: the online model and the *r-hop model*. The $r$-hop model applies to domination problems in which rather than requiring each node in the graph to be served (dominated) by either itself or an adjacent node, it can be served by either itself or any node $r$ hops away from it, that is, any node between which there are at most $r$ edges. In the online version, the entire graph is known in advance; nodes that need to be dominated are revealed over time, and once revealed, they need to be served forever.

The *Online r-hop Connected Dominating Set* problem (r-hop OCDS) belongs to this model. We introduce it in this paper (in Sect. 6), and give a deterministic $\mathcal{O}(2r \log^3 n)$-competitive algorithm, where $n$ is the number of nodes. r-hop OCDS has been studied only in the offline setting; Coelho *et al.* [20] gave inapproximability results for the problem in some special graph classes.

## 2.3 The Online Leasing Model

The *online leasing model* is a generalization of the online model, in which there is a fixed number of lease types, each with a duration and cost; when a resource is leased, it can be used only during its lease duration. If we set the number of lease types to 1 and lease length to infinity, we get the classical online model. The leasing model applies to problems in which requests need to be served for a specific duration, rather than forever. That is, a request arrives at a given time step and needs to be served on that time step only.

The *Online Set Cover Leasing problem* (OSCL) and the *Online Connected Dominating Set Leasing* problem (OCDSL) belong to this model.

– We propose (in Sect. 7) the first deterministic algorithm for OSCL, with $\mathcal{O}(\log \sigma \log(m\mathcal{L} + 2m\frac{\sigma}{l_1}))$-competitive ratio, where $m$ is the number of subsets, $\mathcal{L}$ is the number of lease types, $\sigma$ is the longest lease length, and $l_1$ is the shortest lease length. The currently best result for OSCL is a randomized algorithm by Abshoff et al. [2], with $\mathcal{O}(\log \sigma \log(m\mathcal{L}))$-competitive ratio.
– We introduce OCDSL in this paper (in Sect. 8), and give a deterministic $\mathcal{O}\Big((\sigma + 1)\log \sigma \log(n\mathcal{L} + 2n\frac{\sigma}{l_1}) + \mathcal{L}\log n\Big)$-competitive algorithm, where $n$ is the number of nodes, $\mathcal{L}$ is the number of lease types, $\sigma$ is the longest lease length, and $l_1$ is the shortest lease length.

### 2.4   The Online r-hop Leasing Model

The *online r-hop leasing model* is a combination of the online $r$-hop model and the online leasing model. It applies to domination problems which have the flexibility of serving requests by nodes at most $r$-hops away, the requirement to serve demands only at their arrival time step, and the possibility to lease nodes rather than buying them.

The *Online r-hop Connected Dominating Set Leasing* problem ($r$-hop OCDSL) belongs to this model. We introduce it in this paper (in Sect. 9), and give a deterministic $\mathcal{O}\Big(\mathcal{L}(1 + \sigma(2r - 1)) \log \sigma \log(n\mathcal{L} + 2n\frac{\sigma}{l_1}) \log n\Big)$-competitive algorithm, where $n$ is the number of nodes, $\mathcal{L}$ is the number of lease types, $\sigma$ is the longest lease length, and $l_1$ is the shortest lease length.

## 3   Social Network Applications

Nowadays, people meet virtually more than ever, thanks to social media. The latter has become an advertisement tool for many companies. What better way to promote a product than through influencers? Advertisers are now able to customize the audience they want to target based on their behavior such as their purchasing patterns, the device they use, and other activities. Social media has also been shaping the opinions of people in terms of their social, political, and religious orientation. There has been a vast number of literature about social networks from various perspectives [11,31].

Rather than reaching out all potential customers to advertise a certain product, a company might be able to address just a few of them yet influence all the others and reduce its costs. The question is which people to address so that all potential customers are influenced and the number of influencers is minimized? As the size of social networks grows, the problem becomes more complex. It can be formalized as the *Online Connected Dominating Set* problem (OCDS). Potential customers appear over time; these can be identified based on, for instance, their online activities, searches, and recent purchases. The goal is to choose a smallest possible group of these potential customers, such that each potential customer is either in this group or has a friend in the group. Having a *connected* group would further improve the costs of the company since it will not have to advertise to each group member but rather rely on the group members to tell each other since they are connected.

In certain situations, people are influenced by friends of friends too. Depending on how much the company is willing to pay for advertisement, it may assume a potential customer can be influenced by a friend who is not more than $r$ hops far. For example, for $r = 2$, a potential customer is either in the selected connected group of influencers, or is influenced by a friend, or is influenced by a friend of a friend. This is formalized as the *Online r-hop Connected Dominating Set* problem (r-hop OCDS).

Often, marketers are able to make changes with respect to their audience, along with the nature, duration, and budget of the ads based on its evaluation, as is the case with Facebook, one of the most popular networks worldwide as of today [1]. Due to these dynamics, the groups of influencers are expected to change over time. That is, rather than choosing people to influence others forever, we assume they are chosen for a period of time and need to be chosen again in order to continue being influencers. This can be described as the *online leasing model*, that contains the *Online Connected Dominating Set leasing* problem (OCDSL), the *Online r-hop Connected Dominating Set Leasing* problem (r-hop OCDSL), and the *Online Set Cover Leasing* problem (OSCL), in which the group may not be connected.

## 4 State-of-the-Art

In this section, we give an overview of works related to Online Connected Dominating Set, Online Set Cover, and Leasing problems.

### 4.1 Online Connected Dominating Set Problems

Connected Dominating Set problems have been intensively studied in the offline setting [28,46]. Only few works have considered the online setting. Boyar *et al.* [17] studied an online variant of the *Connected Dominating Set* problem (CDS), in which the input graph is given in advance. The latter is restricted to a unit disk graph, tree, or bounded degree graph, and in each step, nodes are either inserted or deleted. Boyar *et al.* [17] showed that a simple greedy approach attains a $(1 + \frac{1}{\mathcal{OPT}})$-competitive ratio in trees, where $\mathcal{OPT}$ is the cost of the optimal offline solution, an $(8 + \epsilon)$-competitive ratio in unit disk graphs, for arbitrary small $\epsilon > 0$, and $b$-competitive ratio in $b$-bounded degree graphs. In a previous work [30], we introduced the *Online Connected Dominating Set* problem (OCDS), an online variant of CDS, in which the input graph is given in advance. We gave an $\mathcal{O}(\log^2 n)$-competitive randomized algorithm for OCDS in general graphs, where $n$ is the number of nodes. We also ran a simulation study to evaluate the performance of the algorithm in modern robotic warehouses, in which the topology of a warehouse was modeled as a geometric graph.

### 4.2 Online Set Cover Problems

The online variant of the *Set Cover* problem was introduced by Alon *et al.* [8]. They proposed a deterministic $\mathcal{O}(\log m \log n)$-competitive algorithm and showed a nearly matching $\Omega(\log m \log n / (\log \log m + \log \log n))$ deterministic lower bound for many values of $m$ and $n$, where $m$ is the number of subsets and $n$ is the number of elements.

Later, Korman [36] proved a randomized lower bound of $\Omega(\log m \log n)$ for the problem. For the unweighted case where costs are uniform, Alon *et al.* [8] proposed an $\mathcal{O}(\log n \log d)$ competitive ratio. The latter was later improved to $\mathcal{O}(\log(n/Opt) \log d)$ by Buchbinder *et al.* [18], where $Opt$ is the optimal offline solution and $d$ is the maximum number of sets an element belongs to.

### 4.3  Leasing Problems

The leasing model was introduced by Meyerson [43] with the *Parking Permit* problem (PP). Meyerson [43] gave deterministic $\mathcal{O}(\mathcal{L})$-competitive and randomized $\mathcal{O}(\log \mathcal{L})$-competitive algorithms along with matching lower bounds for PP. He also introduced the leasing variant of the *Steiner Forest* problem, for which he proposed a randomized $\mathcal{O}(\log n \log \mathcal{L})$ competitive algorithm, where $n$ is the number of nodes, and $\mathcal{L}$ is the number of lease types. Inspired by Meyerson's work [43], Anthony and Gupta [10] introduced the leasing variants of *Facility Location*, *Set Cover*, and *Steiner Tree*. They showed an interesting connection between infrastructure leasing problems and stochastic optimization problems that leads to approximation algorithms for these variants. Nagarajan and Williamson [45] gave an $\mathcal{O}(\mathcal{L} \log n)$-competitive algorithm for the leasing variant of the *Facility Location* problem, where $n$ is the number of clients. Kling *et al.* [35] extended the latter result by giving an $\mathcal{O}(\sigma \log \sigma)$-competitive algorithm, where $\sigma$ is the maximum lease length. Abshoff *et al.* [2] gave an online randomized algorithm for the leasing variant of *Set Cover*, with $\mathcal{O}(\log(m\mathcal{L}) \log \sigma)$-competitive ratio and improved previous results for other online variants of *Set Cover*. Bienkowski *et al.* [16] proposed a deterministic algorithm that has an $\mathcal{O}(\mathcal{L} \log k)$-competitive ratio for the leasing variant of *Steiner Tree*, where $k$ is the number of terminals. Markarian *et al.* [42] proposed optimal and near-optimal online algorithms for the leasing variants of *Vertex Cover* and *non-metric Facility Location*, respectively. A number of extensions to the *Parking Permit* problem were also studied. Li *et al.* [37] introduced deadlines to the leasing model of Meyerson. Hu *et al.* [32] introduced a two-dimensional variant of the *Parking Permit* problem in which lease types have lengths *and* capacities. Feldkord *et al.* [25] introduced fluctuations to lease prices by considering different models in which lease prices change over time. S. de Lima *et al.* [38] proposed a generalization in which multiple permits can serve an arbitrary demand at a given time step.

## 5   The Online Model: OCDS

In this section, we first define the *Online Connected Dominating Set* problem (OCDS). Then, we introduce an online deterministic algorithm for OCDS and analyze its competitive ratio.

### 5.1  Definition of OCDS and Preliminaries

**Definition 1.** *(OCDS) Given a connected graph $G = (V, E)$ and a sequence of disjoint subsets of $V$ arriving over time. A subset $S$ of $V$ serves as a connected dominating set of a given subset $D$ of $V$ if every node in $D$ is either in $S$ or has an adjacent node in $S$,* *and the subgraph induced by $S$ is connected in $G$. Each step, a subset of $V$ arrives and needs to be served by a connected dominating set of $G$. OCDS asks to grow a connected dominating set of minimum number of nodes.*

*How to Find a Minimal Dominating Set Online?* A *dominating set* of a subset $D$ is a subset $DS$ of nodes such that each node in $D$ is either in $DS$ or has an adjacent node in $DS$. $DS$ is *minimal* if no proper subset of $DS$ is a dominating set of $D$. A minimal dominating set can be constructed online using the online deterministic algorithm by Alon *et al.* for the *Online Set Cover* problem (OSC) [8], the online variant of the classical *Set Cover* problem. A *Set Cover* instance is formed by making each node an element, and corresponding each node to a set that contains the node itself, along with its adjacent nodes. Alon *et al.* [8] gave a deterministic $\mathcal{O}(\log m \log n)$-competitive algorithm for OSC, where $m$ is the number of sets and $n$ is the number of elements.

*How to Find a Steiner Tree Online?* A *Steiner tree* of a subset $D$ is a tree connecting each node in $D$ to a given root $s$. A Steiner tree can be constructed online using the online deterministic $\mathcal{O}(\log n)$-competitive algorithm by Berman *et al.* [15]. The *Steiner tree* problem studied by Berman *et al.* [15] is for edge-weighted graphs and the algorithmic cost is measured by adding the costs of all edges outputted by the online algorithm. Our model in this paper assumes no weights on the nodes, and hence the competitive ratio given by Berman *et al.* for edge-weighted graphs carries over to our graph model in this paper. To see this, assume we are given a graph $G$ with a weight of 1 on all edges and all nodes, and a set of terminals that need to be connected. Let $Opt_e$ be the cost of an optimal Steiner tree $T$ measured by counting the edges in $T$. Let $Opt_n$ be the cost of an optimal Steiner tree $T'$ measured by counting the nodes in $T'$. We have that $Opt_e = Opt_n + 1$. The proof is straightforward, by contradiction. Assume $Opt_e > Opt_n + 1$. We can construct a tree which has an edge cost lower than that of $T$: the tree $T'$ with edge cost $Opt_n + 1$, and this contradicts the fact that $T$ is an optimal Steiner tree. Now assume $Opt_e < Opt_n + 1$. We can construct a tree which has a node cost lower than that of $T'$: the tree $T$ with node cost $Opt_e - 1$, and this contradicts the fact that $T'$ is an optimal Steiner tree. This would not have been the case had there been non-uniform weights on the nodes since the node-weighted variant of the *Steiner tree* problem generalizes the edge-weighted variant by replacing each edge by a node with the corresponding edge cost. Moreover, the node-weighted variant of the *Steiner tree* problem generalizes the *Online Set Cover* problem which has a lower bound of $\Omega(\log m \log n)$ [36] on its competitive ratio.

### 5.2  Online Algorithm for OCDS

The algorithm assigns, at the first time step, any of the nodes purchased by the algorithm, as a root node $s$. At time step $t$:

**Input:** $G = (V, E)$, subset $D_t$ of $V$
**Output:** A connected dominating set $CDS_t$ of $D_t$

1. Find a minimal dominating set $DS_t$ of $D_t$.

2. Assign to each node in $DS_t$ a *connecting* node, that is any adjacent node from the set $D_t$. If $t = 1$, assign any of the nodes in $DS_t$ as a root node $s$.
3. Find a Steiner tree that connects all connecting nodes to $s$. Add all the nodes in this tree including the nodes in $DS_t$ and their connecting nodes to $CDS_t$.

### 5.3   Competitive Analysis

**Lower Bound.** OCDS has a lower bound of $\Omega(\log^2 n)$, where $n$ is the number of nodes, resulting from Korman's lower bound of $\Omega(\log m \log n)$ for OSC [36], where $m$ is the number of subsets and $n$ is the number of elements.

**Upper Bound.** Let $Opt$ be the cost of an optimal solution $Opt_I$ of an instance $I$ of OCDS. Let $C1$, $C2$, and $C3$ be the cost of the algorithm in the three steps, respectively. The first step of the algorithm constructs online a minimal dominating set. Let $Opt_{DS}$ be the cost of a minimum dominating set of $I$. Note that $Opt_I$ is a dominating set of $I$. Hence, Alon *et al.*'s [8] deterministic algorithm yields:

$$C1 \le \log^2 n \cdot Opt_{DS} \le \log^2 n \cdot Opt$$

The second step adds at most one node for each node bought in the first step. Hence we have that:

$$C2 \le C1$$

As for the third step, $Opt_I$ is a Steiner tree for the connecting nodes bought in the second step, since all connecting nodes belong to the set of nodes that need to be served and $Opt_I$ serves as a connected dominating set of these nodes. Let $Opt_{St}$ be the cost of a minimum Steiner tree of these connecting nodes. Since Berman *et al.*'s [15] algorithm has an $\mathcal{O}(\log n)$-competitive ratio, we conclude the following:

$$C3 \le \log n \cdot Opt_{St} \le \log n \cdot Opt$$

The total cost of the algorithm is then upper bounded by: $C1 + C2 + C3 = (2 \cdot \log^2 n + \log n) \cdot Opt$ and the theorem below follows.

**Theorem 1.** *There is an asymptotically optimal $\mathcal{O}(\log^2 n)$-competitive deterministic algorithm for the Online Connected Dominating Set problem, where $n$ is the number of nodes.*

## 6   The Online r-hop Model: r-hop OCDS

In this section, we first define the *Online r-hop Connected Dominating Set* problem (r-hop OCDS). Then, we introduce an online deterministic algorithm for r-hop OCDS and analyze its competitive ratio.

### 6.1   Definition of r-hop OCDS and Preliminaries

**Definition 2.** *(r-hop OCDS) Given a connected graph $G = (V, E)$, a positive integer r, and a sequence of disjoint subsets of V arriving over time. A subset S of V serves as an r-hop connected dominating set of a given subset D of V if for every node v in D, there is a vertex u in S such that there are at most r hops (edges) between v and u in G, and the subgraph induced by S is connected in G. Each step, a subset of V arrives and needs to be served by an r-hop connected dominating set of G. r-hop OCDS asks to grow an r-hop connected dominating set of minimum number of nodes. OCDS is equivalent to r-hop OCDS with $r = 1$.*

*How to Find a Minimal r-hop Dominating Set?*   Given a graph $G = (V, E)$ and a positive integer $r$. A subset $DS$ of $V$ is an *r-hop dominating set* of a given subset $D$ of $V$ if for every node $v$ in $D$, there is a vertex $u$ in $DS$ such that there are at most $r$ hops between $v$ and $u$ in $G$. $DS$ is *minimal* if no proper subset of $DS$ is an *r-hop dominating set* of $D$. We can transform an *r-hop dominating set* instance into a *Set Cover* instance by making each node an element, and corresponding each node to a set that contains the node itself, along with all nodes that are at most $r$ hops away from it. Hence, we can construct a minimal *r-hop dominating set* by running the online deterministic algorithm by Alon *et al.* for the *Online Set Cover* problem (OSC) [8].

*How to Find a Steiner Tree?*   A Steiner tree can be constructed online, as in Sect. 5.1 of OCDS, using the online deterministic $\mathcal{O}(\log n)$-competitive algorithm by Berman *et al.* [15].

### 6.2   Online Algorithm for r-hop OCDS

The algorithm assigns, at the first time step, any of the nodes purchased by the algorithm, as a root node $s$. At time step $t$:

**Input:** $G = (V, E)$, subset $D_t$ of $V$
**Output:** An $r$-hop connected dominating set $rCDS_t$ of $D_t$

1. Find a minimal $r$-hop dominating set $rDS_t$ of $D_t$. If $t = 1$, assign any of the nodes in $rDS_t$ as a root node $s$.
2. Find a Steiner tree that connects all nodes in $rDS_t$ to $s$. Add all the nodes in this tree including the nodes in $rDS_t$ to $rCDS_t$.

### 6.3   Competitive Analysis

**Lower Bound.** The only lower bound for r-hop OCDS is the one for OCDS, $\Omega(\log^2 n)$, where $n$ is the number of nodes.

**Upper Bound.** Let $Opt$ be the cost of an optimal solution $Opt_I$ of an instance $I$ of r-hop OCDS. Let $C1$ and $C2$ be the cost of the algorithm in the two steps, respectively. The first step of the algorithm constructs online a minimal $r$-hop dominating set. Let $Opt_{DS}$ be the cost of a minimum $r$-hop dominating set of $I$. Note that $Opt_I$ is an $r$-hop dominating set of $I$. Hence, Alon *et al.*'s [8] deterministic algorithm yields:

$$C1 \le \log^2 n \cdot Opt_{DS} \le \log^2 n \cdot Opt$$

For the second step, we compare the cost of the algorithm to that of a Steiner tree $S'$ which we construct as follows. We refer to the nodes bought in the first step as terminals. $S'$ will connect all the terminals to each other. We add to $S'$ all the nodes in $Opt_I$ and all the terminals. We also add to $S'$ additional nodes that connect each terminal node to $Opt_I$ through a path containing a node from the demand set. For each terminal node, we need at most $2(r-1) + 1$ nodes to connect to $Opt_I$, since $Opt_I$ is an $r$-hop connected dominating set of the demand set. The cost of $S'$ is upper bounded by: $Opt + C1 + (2r - 1) \cdot C1$. Note that the cost of $S'$ is an upper bound to the cost of an optimal Steiner tree for the same set of terminals. Let $Opt_{St}$ be the cost of a minimum Steiner tree connecting the terminals. Since Berman *et al.*'s [15] algorithm has an $\mathcal{O}(\log n)$-competitive ratio, we conclude the following:

$$C2 \le \log n \cdot Opt_{St} \le \log n \cdot (Opt + 2r \cdot C1) \le (\log n + 2r \log^3 n) \cdot Opt$$

Hence, the following theorem follows.

**Theorem 2.** *There is a deterministic $\mathcal{O}(2r \cdot \log^3 n)$-competitive algorithm for the Online r-hop Connected Dominating Set problem, where $n$ is the number of nodes.*

## 7  The Online Leasing Model: OSCL

In this section, we first define the *Online Set Cover Leasing* problem (OSCL). Then, we introduce an online deterministic algorithm for OSCL and analyze its competitive ratio.

### 7.1  Definition of OSCL and Preliminaries

**Definition 3.** *(OSCL) Given a universe $\mathcal{U}$ of elements ($|\mathcal{U}| = n$), a collection $\mathcal{S}$ of subsets of $\mathcal{U}$ ($|\mathcal{S}| = m$), and a set of $\mathcal{L}$ different lease types, each characterized by a duration and cost. A subset can be leased using lease type $l$ for cost $c_l$ and remains active for $d_l$ time steps. Each time step $t$, an element $e \in \mathcal{U}$ arrives and there needs to be a subset $S \in \mathcal{S}$ active at time $t$ such that $e \in S$. OSCL asks to minimize the total leasing costs.*

*How to Simplify the Problem without Losing Asymptotically?* We assume a simplified configuration on the leases. The latter has been similarly defined by Meyerson for the *Parking Permit* problem [43], who showed that by assuming this, one loses only a constant factor in the competitive ratio. A similar argument can be easily made for OSCL, as was the case for all generalizations of the *Parking Permit* problem [2, 16, 45]. The configuration is defined as follows.

**Definition 4.** *(Lease Configuration) Leases of type $l$ only start at times multiple of $d_l$, where $d_l$ is the length of lease type $l$. Moreover, all lease lengths are power of two.*

*Which Online Covering Algorithm Is Needed?* Our algorithm for OSCL is based on running Alon *et al.*'s [8] deterministic algorithm for the *Online Set Cover* problem (the weighted case), which constructs a fractional solution that is rounded online into an integral deterministic solution. Alon *et al.*'s algorithm has an $\mathcal{O}(\log m \log n)$-competitive ratio and requires the knowledge of the set cover instance to make it deterministic. What is unknown to the algorithm is the order and subset of arriving elements. We will transform an instance $\alpha$ of OSCL into an instance $\alpha'$ of the *Online Set Cover* problem and run Alon *et al.*'s deterministic algorithm on $\alpha'$. An instance of the *Online Set Cover* problem consists of a universe of elements and a collection of subsets of the universe - an element of the universe arrives in each step. The algorithm needs to purchase subsets such that each arriving element is covered, upon its arrival, by one of these subsets, while minimizing the total costs of subsets. The algorithm may end up covering elements that never arrive.

### 7.2  Online Algorithm for OSCL

Suppose the algorithm is given a universe $\mathcal{U}$ of elements and a collection $\mathcal{S}$ of subsets of $\mathcal{U}$. If there is one lease type, of infinite lease length ($\mathcal{L} = 1$), we have exactly an instance of the *Online Set Cover* problem and so Alon *et al.*'s [8] deterministic algorithm would solve it. Otherwise, we do the following - we represent each element $e \in \mathcal{U}$ by $n$ pairs, one for each of the at most $n$ potential time steps at which $e$ can arrive. We let pair $(e, t)$ represent element $e$ at time step $t$. We denote by $\mathcal{N}$ the collection of all these pairs. A subset $S \in \mathcal{S}$ can be leased using lease type $l$ for cost $c_l$ and remains *active* for $d_l$ time steps. We represent subset $S$ of lease type $l$ at time $t$ as a triplet $(S, l, t)$. We denote by $\mathcal{M}$ the collection of all these triplets. We now construct an instance of the *Online Set Cover* problem with $\mathcal{N}$ and $\mathcal{M}$ being the collection of elements and of subsets, respectively. Pair $(e, t)$ can be covered by triplet $(S, l, t')$ if $e \in S$ and $t \in [t', t' + d_l]$. When an element arrives at time $t$, pair $(e, t)$ is given as input to the *Online Set Cover* instance for step $t$. Note that each element $e \in \mathcal{U}$ arrives only once. An algorithm for the *Online Set Cover* problem will ensure that $e's$ corresponding pair at the time it arrives is covered. Moreover it will ignore (not necessarily cover) all other pairs corresponding to the other time steps and this is equivalent to having elements that never arrive in an *Online Set Cover* instance. Hence, running Alon *et al.*'s [8] algorithm will yield a feasible deterministic solution for OSCL.

### 7.3  Competitive Analysis

**Lower Bound.** OSCL has a lower bound of $\Omega(\log m \log n + L)$ resulting from the $\Omega(\log m \log n)$ lower bound for OSC [36], where $m$ is the number of subsets and $n$ is the number of elements, and the $\Omega(L)$ lower bound for the *Parking Permit* problem [43], where $L$ is the number of lease types.

**Upper Bound.** We fix any interval $I$ of length $\sigma$ and show that the algorithm would be $\mathcal{O}(\log \sigma \log(m\mathcal{L} + 2m\frac{\sigma}{l_1}))$-competitive if this interval were the entire input, where $l_1$ is the length of the shortest lease, $\sigma$ is the length of the longest lease, $\mathcal{L}$ is the number of lease types, and $m$ is the number of subsets. Since all leases including the ones in the optimal solution end at the end of $I$ due to the lease configuration defined earlier, this would imply that the algorithm has an $\mathcal{O}(\log \sigma \log(m\mathcal{L} + 2m\frac{\sigma}{l_1}))$-competitive ratio. Note that there are at most $\sigma$ elements over $I$, since at most one element arrives in each time step. The competitive ratio $\mathcal{O}(\log |\mathcal{M}| \log |\mathcal{N}|)$ of the algorithm follows directly by setting the number of elements and subsets to $|\mathcal{N}|$ and $|\mathcal{M}|$, respectively. Now, we have that $|\mathcal{N}| = \sigma^2$ since there are $\sigma^2$ pairs in total. Next, we give an upper bound to $|\mathcal{M}|$ over $I$.

$$|\mathcal{M}| \leq m \cdot \left( \sum_{j=1}^{\mathcal{L}} \left\lceil \frac{\sigma}{l_j} \right\rceil \right)$$

Since $l_j$s are increasing and powers of two, we conclude that the sum above can be upper bounded by the sum of a geometric series with a ratio of $1/2$.

$$\sum_{j=1}^{\mathcal{L}} \left\lceil \frac{\sigma}{l_j} \right\rceil \leq \mathcal{L} + \sigma \left[ \frac{1}{l_1} \left( \frac{1 - (1/2)^\sigma}{1 - 1/2} \right) \right] = \mathcal{L} + \sigma \left[ \frac{2}{l_1} \left( 1 - (1/2)^{\mathcal{L}} \right) \right]$$

Since $\mathcal{L} \geq 1$, we have:

$$\mathcal{L} + \sigma \left[ \frac{2}{l_1} \left( 1 - (1/2)^{\mathcal{L}} \right) \right] \leq \mathcal{L} + \frac{2\sigma}{l_1}.$$

Therefore, $|\mathcal{M}| \leq m \cdot (\mathcal{L} + \frac{2\sigma}{l_1})$, and the theorem below follows.    □

**Theorem 3.** *There is a deterministic $\mathcal{O}(\log \sigma \log(m\mathcal{L} + 2m\frac{\sigma}{l_1}))$-competitive algorithm for the Online Set Cover Leasing problem, where $m$ is the number of subsets, $\mathcal{L}$ is the number of lease types, $\sigma$ is the longest lease length, and $l_1$ is the shortest lease length.*

**Remark.** Note that, in the $\Omega(L)$ lower bound formulation for the *Parking Permit* problem [43], the longest lease length $\sigma$ is exponential in $L$. This explains why the upper bound obtained for OSCL is achievable.

## 8   The Online Leasing Model: OCDSL

In this section, we first define the *Online Connected Dominating Set Leasing* problem (OCDSL). Then, we introduce an online deterministic algorithm for OCDSL and analyze its competitive ratio.

### 8.1   Definition of OCDSL and Preliminaries

**Definition 5.** *(OCDSL) Given a connected graph $G = (V, E)$, a sequence of disjoint subsets of $V$ arriving over time, and a set of $\mathcal{L}$ different lease types, each characterized* *by a duration and cost. A node can be leased using lease type $l$ for cost $c_l$ and remains active for $d_l$ time steps. A subset $S$ of nodes of $V$ serves as a connected dominating set of a given subset $D$ of $V$ if every node in $D$ is either in $S$ or has an adjacent node in $S$, and the subgraph induced by $S$ is connected in $G$. Each time step $t$, a subset of $V$ arrives and needs to be served by a connected dominating set of nodes active at time $t$. OCDSL asks to grow a connected dominating set with minimum leasing costs.*

OCDS is equivalent to OCDSL with one lease type ($L = 1$) of infinite length. Note that in both OCDS and OCDSL, the algorithm ends up purchasing (leasing) nodes that form one connected subgraph - the difference is that in OCDSL, at a certain time step $t$, only the currently active nodes needed to serve the nodes given at time $t$, are connected by nodes active at time $t$, to at least one of the previously leased nodes, thus maintaining one single connected subgraph.

*Which Online Leasing Algorithms Are Needed?* We assume the lease configuration introduced earlier in Definition 4. We use two leasing algorithms:

- our online deterministic algorithm for OSCL presented in Sect. 7.2. Recall that it has an $\mathcal{O}(\log \sigma \log(m\mathcal{L} + 2m\frac{\sigma}{l_1}))$-competitive ratio, where $m$ is the number of subsets, $\mathcal{L}$ is the number of lease types, $\sigma$ is the longest lease length, and $l_1$ is the shortest lease length.
- the online deterministic algorithm for the *Online Steiner Tree Leasing* problem (OSTL) by Bienkowski *et al.* [16]. Given a connected graph $G = (V, E)$, a root node $s$, a sequence of nodes of $V$ (called *terminals*) arriving over time, and a set of $\mathcal{L}$ different lease types, each characterized by a duration and cost. An edge can be leased using lease type $l$ for cost $c_l$ and remains active for $d_l$ time steps. Each step $t$, a node arrives and needs to be connected to $s$ through a path of edges active at time $t$. OSTL asks to minimize the total leasing costs. The algorithm by Bienkowski *et al.* [16] has an $\mathcal{O}(\mathcal{L} \log k)$-competitive ratio, where $k$ is the number of terminals.

## 8.2   Online Algorithm for OCDSL

The algorithm assigns, at the first time step, any of the nodes leased by the algorithm, as a root node $s$. At time step $t$:

**Input:** $G = (V, E)$, subset $D_t$ of $V$
**Output:** A set of leased nodes that form a connected dominating set of $D_t$

1. Lease a set $DS_t$ of nodes that form a minimal dominating set of $D_t$.
2. Assign to each node in $DS_t$ a *connecting* node, that is any adjacent node from the set $D_t$. Buy the cheapest lease for each of these connecting nodes. If $t = 1$, assign any of the nodes in $DS_t$ as a root node $s$.
3. Lease a set of nodes that connect all connecting nodes to $s$.

To find a set of leased nodes that form a *minimal dominating set* of a subset $D_t$, we run our deterministic algorithm for OSCL (see Sect. 7.2). An *Online Set Cover Leasing* instance is formed by making each node an element, and corresponding each node to a set that contains the node itself, along with its adjacent nodes - sets are leased with $\mathcal{L}$ different lease types.

To find a set of leased nodes that connect a subset of nodes to $s$, we run the deterministic algorithm for the *Online Steiner Tree Leasing* problem (OSTL) by Bienkowski *et al.* [16]. The problem studied by Bienkowski *et al.* [16] is for edge-weighted graphs and the algorithmic cost is measured by adding the leasing costs of the edges and not the nodes. Our model in this paper assumes no weights on the nodes, and hence the competitive ratio given by Bienkowski *et al.* [16] for edge-weighted graphs carries over to our graph model in this paper. This would not have been the case had there been non-uniform weights on the nodes since the node-weighted variant of the *Online Steiner Tree Leasing* problem generalizes the edge-weighted variant. Hence, whenever the algorithm for OSTL leases an edge $(u, v)$ at time $t$ with lease type $l$, we lease both $u$ and $v$ at the same time $t$ with the same lease type $l$ and hence the cost will only double.

### 8.3 Competitive Analysis

**Lower Bound.** Since OCDSL generalizes OSCL, $\Omega(\log^2 n + L)$ is a lower bound for OCDSL, where $n$ is the number of nodes and $L$ is the number of lease types.

**Upper Bound.** Let $Opt$ be the cost of an optimal solution $Opt_I$ of an instance $I$ of OCDSL. Let $C1$, $C2$, and $C3$ be the cost of the algorithm in the three steps, respectively. The first step of the algorithm constructs online a minimal dominating set. Let $Opt_{DS}$ be the cost of a minimum dominating set of $I$. Note that $Opt_I$ is a dominating set of $I$. Hence, our deterministic $\mathcal{O}(\log \sigma \log(m\mathcal{L} + 2m\frac{\sigma}{l_1}))$-competitive algorithm for *Online Set Cover Leasing* yields:

$$C1 \leq \log \sigma \log(n\mathcal{L} + 2n\frac{\sigma}{l_1}) \cdot Opt_{DS} \leq \log \sigma \log(n\mathcal{L} + 2n\frac{\sigma}{l_1}) \cdot Opt$$

In the second step, the algorithm buys the cheapest lease type, multiple times. We show that the cost of all these leases is upper bounded by $\sigma \cdot C1$. We denote by $DS$ the set of all leases purchased by the algorithm in the first step. Let $c_1$ denote the cost of the cheapest lease. The cardinality of $DS$ can be upper bounded by $C1/c_1$, since each lease in $DS$ has a cost of at least $c_1$. Fix any node lease $i$ in $DS$ of lease type $l$. The algorithm purchases at most $d_l$ connecting nodes corresponding to $i$, each of cost $c_1$, where $d_l$ is the length of lease type $l$. This is because $i$ may be assigned by the algorithm multiple times, during its lease duration. Hence, the total cost of all connecting nodes purchased by the algorithm is at most $c_1 \cdot C1/c_1 \cdot \sigma$, where $\sigma$ is the longest lease length. Therefore,

$$C2 \leq \sigma \cdot C1$$

As for the third step, $Opt_I$ is a Steiner tree for the connecting nodes bought in the second step, since all connecting nodes belong to the set of nodes that need to be served and $Opt_I$ serves as a connected dominating set of these nodes. Let $Opt_{St}$ be the cost of a minimum Steiner tree of these connecting nodes. Since Bienkowski *et al.*'s [16] algorithm has an $\mathcal{O}(\mathcal{L} \log k)$-competitive ratio, we conclude the following:

$$C3 \leq \mathcal{L} \cdot \log n \cdot Opt_{St} \leq \mathcal{L} \cdot \log n \cdot Opt$$

The total cost of the algorithm is then upper bounded by: $C1 + C2 + C3 = ((\sigma + 1) \cdot \log \sigma \log(n\mathcal{L} + 2n\frac{\sigma}{l_1}) + \mathcal{L} \cdot \log n) \cdot Opt$ and the theorem below follows.

**Theorem 4.** *There is a deterministic* $\mathcal{O}\left((\sigma+1)\cdot\log\sigma\log(n\mathcal{L}+2n\frac{\sigma}{l_1})+\mathcal{L}\cdot\log n\right)$ *competitive algorithm for the Online Connected Dominating Set Leasing problem, where* $n$ *is the number of nodes,* $\mathcal{L}$ *is the number of lease types,* $\sigma$ *is the longest lease length, and* $l_1$ *is the shortest lease length.*

## 9   The Online r-hop Leasing Model: r-hop OCDSL

In this section, we first define the *Online r-hop Connected Dominating Set Leasing* problem (r-hop OCDSL). Then, we introduce an online deterministic algorithm for r-hop OCDSL and analyze its competitive ratio.

### 9.1   Definition of r-hop OCDSL and Preliminaries

**Definition 6.** *Given a connected graph* $G = (V, E)$, *a positive integer* $r$, *a sequence of disjoint subsets of* $V$ *arriving over time, and a set of* $\mathcal{L}$ *different lease types, each characterized by a duration and cost. A node can be leased using lease type* $l$ *for cost* $c_l$ *and remains active for* $d_l$ *time steps. A subset* $S$ *of nodes of* $V$ *serves as an* $r$-*hop connected dominating set of a given subset* $D$ *of* $V$ *if for every node* $v$ *in* $D$, *there is a vertex* $u$ *in* $S$ *such that there are at most* $r$ *hops between* $v$ *and* $u$ *in* $G$, *and the subgraph induced by* $S$ *is connected in* $G$. *Each time step* $t$, *a subset of* $V$ *arrives and needs to be served by an* $r$-*hop connected dominating set of nodes active at time* $t$. $r$-*hop OCDSL asks to grow an* $r$-*hop connected dominating set with minimum leasing costs.*

OCDSL is equivalent to r-hop OCDSL for $r = 1$.

*Which Online Leasing Algorithms Are Needed?*  We assume the lease configuration introduced earlier in Definition 4. We use two leasing algorithms:

- our online deterministic algorithm for OSCL presented in Sect. 7.2. Recall that it has an $\mathcal{O}(\log \sigma \log(m\mathcal{L} + 2m\frac{\sigma}{l_1}))$-competitive ratio, where $m$ is the number of subsets, $\mathcal{L}$ is the number of lease types, $\sigma$ is the longest lease length, and $l_1$ is the shortest lease length.
- the online deterministic algorithm for the *Online Steiner Tree Leasing* problem (OSTL) by Bienkowski *et al.* [16]. The algorithm by Bienkowski *et al.* [16] has an $\mathcal{O}(\mathcal{L} \log k)$-competitive ratio, where $k$ is the number of terminals.

### 9.2   Online Algorithm for r-hop OCDSL

The algorithm assigns, at the first time step, any of the nodes leased by the algorithm, as a root node $s$. At time step $t$:

**Input:** $G = (V, E)$ and subset $D_t$ of $V$
**Output:** A set of leased nodes that form $r$-hop connected dominating set of $D_t$

1. Lease a set $rDS_t$ of nodes that form a minimal $r$-hop dominating set of $D_t$. If $t = 1$, assign any of the nodes in $rDS_t$ as a root node $s$.
2. Lease a set of nodes that connect all nodes in $rDS_t$ to $s$.

To find a set of leased nodes that form a *minimal r-hop dominating set* of a subset $D_t$, we run our deterministic algorithm for OSCL presented in Sect. 7.2. An Online Set Cover Leasing instance is formed by making each node an element, and corresponding each node to a set that contains the node itself, along with all nodes that are at most $r$ hops away from it; sets are leased with $\mathcal{L}$ different lease types.

To find a set of leased nodes that connect a subset of nodes to $s$, we run the deterministic algorithm for the *Online Steiner Tree Leasing* problem (OSTL) by Bienkowski *et al.* [16].

### 9.3   Competitive Analysis

**Lower Bound.** The only lower bound for r-hop OCDSL is the one for OCDSL, $\Omega(\log^2 n + L)$, where $n$ is the number of nodes and $L$ is the number of lease types.

**Upper Bound.** Let $Opt$ be the cost of an optimal solution $Opt_I$ of an instance $I$ of r-hop OCDSL. Let $C1$ and $C2$ be the cost of the algorithm in the two steps, respectively. The first step of the algorithm constructs online a minimal $r$-hop dominating set. Let $Opt_{DS}$ be the cost of a minimum $r$-hop dominating set of $I$. Note that $Opt_I$ is an $r$-hop dominating set of $I$. Hence, our deterministic $\mathcal{O}(\log \sigma \log(m\mathcal{L} + 2m\frac{\sigma}{l_1}))$-competitive algorithm for *Online Set Cover Leasing* yields:

$$C1 \leq \log \sigma \log(n\mathcal{L} + 2n\frac{\sigma}{l_1}) \cdot Opt_{DS} \leq \log \sigma \log(n\mathcal{L} + 2n\frac{\sigma}{l_1}) \cdot Opt$$

For the second step, we compare the leasing cost of the algorithm to that of a Steiner tree $S'$ which we construct as follows. $S'$ is composed of leased nodes that connect, for each time step $t$, the nodes in $rDS_t$ to $s$. We refer to the leases bought in the first step of the algorithm as terminal leases and their corresponding nodes as terminal nodes. The total number of terminal leases can be upper bounded by $C1/c_1$, where $c_1$ is the cost of the shortest lease. We add to $S'$ all the leases in $Opt_I$ and all the terminal leases. We also add to $S'$ additional leases that connect each terminal node to $Opt_I$ through a path containing a node from the demand set. Fix any terminal lease of type $l$ leased at time $t$ with terminal node $u$. At most $2(r-1) + 1$ nodes are needed to connect $u$ to $Opt_I$, since $Opt_I$ is an $r$-hop connected dominating set of the demand set. For each of these

$2(r-1)+1$ nodes, we buy the cheapest lease at multiples times that cover the interval $[t, t + d_l]$. We add these leases to $S'$. The cost of $S'$ will then be upper bounded by: $Opt + C1 + \left((2r - 1) \cdot C1/c_1 \cdot c_1 \cdot \sigma\right)$. Note that the cost of $S'$ is an upper bound to the cost of an optimal Steiner tree for the same set of terminal nodes. Let $Opt_{St}$ be the cost of a minimum Steiner tree connecting the terminal nodes. Since Bienkowski *et al.*'s [16] algorithm has an $\mathcal{O}(\mathcal{L} \log k)$-competitive ratio, we conclude the following:

$$C2 \leq \mathcal{L} \log n \cdot Opt_{St}$$

$$\leq \mathcal{L} \log n \cdot \left(Opt + C1(1 + \sigma(2r - 1))\right)$$

$$\leq \mathcal{L} \log n \cdot \left(Opt + \log \sigma \log(n\mathcal{L} + 2n\frac{\sigma}{l_1}) \cdot Opt \cdot (1 + \sigma(2r - 1)))\right)$$

$$= Opt\left(\mathcal{L} \log n + \log \sigma \log(n\mathcal{L} + 2n\frac{\sigma}{l_1}) \cdot (1 + \sigma(2r - 1)) \cdot \mathcal{L} \log n\right)$$

Hence, the following theorem follows.

**Theorem 5.** *There is a deterministic* $\mathcal{O}\left(\mathcal{L}(1+\sigma(2r-1)) \log \sigma \log(n\mathcal{L}+2n\frac{\sigma}{l_1}) \log n\right)$-*competitive algorithm for the Online r-hop Connected Dominating Set Leasing problem, where $n$ is the number of nodes, $\mathcal{L}$ is the number of lease types, $\sigma$ is the longest lease length, and $l_1$ is the shortest lease length.*

## 10   Concluding Remarks and Future Work

Many real-world application problems have an online nature, making them difficult to handle. The study of *online algorithms* is one way to approach these problems. In this paper, we have particularly targeted domination and covering problems, inspired by social network advertising. There is room for a lot of interesting research.

Proceeding with competitive analysis, one may want to close the gaps between the upper and lower bounds; a summary of the results is given in Table 1 below.

While competitive analysis provides a worst-case guarantee for the performance of the online algorithms, it is always good to understand how these algorithms would perform in actual real-world scenarios. In a previous study for OCDS [30], we implemented the proposed algorithm in a simulated robotic warehouse modeled by geometric graphs. The results were promising as it turned out that the algorithm performs well on average too. Similarly, one may want to understand the average performance of the algorithms proposed in this paper, for instance, in social network applications; an interesting set of social network scenarios has been presented in Sect. 3.

There are a lot of domination problems that have been introduced in various real-world contexts, such as in wireless networks [46]. One alternative is to define these in the problem models proposed in this paper and solve them in these online settings. Another way to go is to study these including the problems addressed in this paper in geometric graphs. There have been many works that resulted in approximation ratios that are dependent on the properties of the geometric graphs [12, 33]; it is not yet explored though whether the same arguments carry over to online settings.

**Table 1.** Summary of asymptotic bounds.

| Problem | Lower bound | Upper bound |
|---|---|---|
| OCDS | $\Omega(\log^2 n)$ | $\mathcal{O}(\log^2 n)$ |
| r-hop OCDS | $\Omega(\log^2 n)$ | $\mathcal{O}(2r \log^3 n)$ |
| OSCL | $\Omega(\log m \log n + L)$ | $\mathcal{O}(\log \sigma \log(m\mathcal{L} + 2m\frac{\sigma}{l_1}))$ |
| OCDSL | $\Omega(\log^2 n + L)$ | $\mathcal{O}\left((\sigma + 1)\log \sigma \log(n\mathcal{L} + 2n\frac{\sigma}{l_1}) + \mathcal{L}\log n\right)$ |
| r-hop OCDSL | $\Omega(\log^2 n + L)$ | $\mathcal{O}\left(\mathcal{L}(1 + \sigma(2r - 1))\log \sigma \log(n\mathcal{L} + 2n\frac{\sigma}{l_1})\log n\right)$ |

# References

1. Facebook business; choose your audience. https://www.facebook.com/business. Accessed 27 May 2018

2. Abshoff, S., Kling, P., Markarian, C., Meyer auf der Heide, F., Pietrzyk, P.: Towards the price of leasing online. J. Comb. Optim. **32**(4), 1197–1216 (2016). https://doi.org/10.1007/s10878-015-9915-5

3. Abu-Khzam, F.N., Heggernes, P.: Enumerating minimal dominating sets in chordal graphs. Inf. Process. Lett. **116**(12), 739–743 (2016). https://doi.org/10.1016/j.ipl.2016.07.002

4. Abu-Khzam, F.N., Lamaa, K.: Efficient heuristic algorithms for positive-influence dominating set in social networks. In: IEEE Conference on Computer Communications Workshops, INFOCOM Workshops 2018, IEEE INFOCOM 2018, Honolulu, HI, USA, 15–19 April 2018, pp. 610–615 (2018). https://doi.org/10.1109/INFOCOMW.2018.8406851

5. Abu-Khzam, F.N., Li, S., Markarian, C., Meyer auf der Heide, F., Podlipyan, P.: On the parameterized parallel complexity and the vertex cover problem. In: Chan, T.-H.H., Li, M., Wang, L. (eds.) COCOA 2016. LNCS, vol. 10043, pp. 477–488. Springer, Cham (2016). https://doi.org/10.1007/978-3-319-48749-6_35

6. Abu-Khzam, F.N., Li, S., Markarian, C., Meyer auf der Heide, F., Podlipyan, P.: Efficient parallel algorithms for parameterized problems. Theor. Comput. Sci. **786**, 2–12 (2019). https://doi.org/10.1016/j.tcs.2018.11.006

7. Abu-Khzam, F.N., Markarian, C., Meyer auf der Heide, F., Schubert, M.: Approximation and heuristic algorithms for computing backbones in asymmetric ad-hoc networks. Theory Comput. Syst. **62**(8), 1673–1689 (2018). https://doi.org/10.1007/s00224-017-9836-z

8. Alon, N., Awerbuch, B., Azar, Y.: The online set cover problem. In: Proceedings of the Thirty-Fifth Annual ACM Symposium on Theory of Computing, STOC 2003, pp. 100–105. ACM, New York (2003). https://doi.org/10.1145/780542.780558

9. Alon, N., Moshkovitz, D., Safra, S.: Algorithmic construction of sets for k-restrictions. ACM Trans. Algorithms **2**(2), 153–177 (2006). https://doi.org/10.1145/1150334.1150336

10. Anthony, B.M., Gupta, A.: Infrastructure leasing problems. In: Fischetti, M., Williamson, D.P. (eds.) IPCO 2007. LNCS, vol. 4513, pp. 424–438. Springer, Heidelberg (2007). https://doi.org/10.1007/978-3-540-72792-7_32

11. Arafeh, M., Ceravolo, P., Mourad, A., Damiani, E.: Sampling online social networks with tailored mining strategies. In: 2019 Sixth International Conference on Social Networks Analysis, Management and Security (SNAMS), pp. 217–222. IEEE (2019)

12. Bai, X., Zhao, D., Bai, S., Wang, Q., Li, W., Mu, D.: Minimum connected dominating sets in heterogeneous 3D wireless ad hoc networks. Ad Hoc Networks **97**, 102023 (2020). https://doi.org/10.1016/j.adhoc.2019.102023. http://www.sciencedirect.com/science/article/pii/S1570870518304074

13. Barman, S., Pal, M., Mondal, S.: An optimal algorithm to find minimum k-hop dominating set of interval graphs. Discret. Math. Algorithms Appl. **11** (2018). https://doi.org/10.1142/S1793830919500162

14. Berger, B., Rompel, J., Shor, P.W.: Efficient NC algorithms for set cover with applications to learning and geometry. J. Comput. Syst. Sci. **49**(3), 454–477 (1994). https://doi.org/10.1016/S0022-0000(05)80068-6. http://www.sciencedirect.com/science/article/pii/S0022000005800686. 30th IEEE Conference on Foundations of Computer Science

15. Berman, P., Coulston, C.: On-line algorithms for Steiner tree problems (extended abstract), pp. 344–353 (1997). https://doi.org/10.1145/258533.258618

16. Bienkowski, M., Kraska, A., Schmidt, P.: A deterministic algorithm for online Steiner tree leasing. In: WADS 2017. LNCS, vol. 10389, pp. 169–180. Springer, Cham (2017). https://doi.org/10.1007/978-3-319-62127-2_15

17. Boyar, J., Eidenbenz, S.J., Favrholdt, L.M., Kotrbčík, M., Larsen, K.S.: Online dominating set. In: 15th Scandinavian Symposium and Workshops on Algorithm Theory, SWAT 2016, Reykjavik, Iceland, 22–24 June 2016, pp. 21:1–21:15 (2016). https://doi.org/10.4230/LIPIcs.SWAT.2016.21

18. Buchbinder, N., Naor, J.: Online primal-dual algorithms for covering and packing problems. In: Brodal, G.S., Leonardi, S. (eds.) ESA 2005. LNCS, vol. 3669, pp. 689–701. Springer, Heidelberg (2005). https://doi.org/10.1007/11561071_61

19. Chvatal, V.: A greedy heuristic for the set-covering problem. Math. Oper. Res. **4**(3), 233–235 (1979). http://www.jstor.org/stable/3689577

20. Coelho, R.S., Moura, P.F.S., Wakabayashi, Y.: The k-hop connected dominating set problem: approximation and hardness. J. Comb. Optim. **34**(4), 1060–1083 (2017). https://doi.org/10.1007/s10878-017-0128-y

21. Daliri Khomami, M.M., Rezvanian, A., Bagherpour, N., Meybodi, M.R.: Minimum positive influence dominating set and its application in influence maximization: a learning automata approach. Appl. Intell. **48**(3), 570–593 (2018). https://doi.org/10.1007/s10489-017-0987-z

22. Du, D.Z., Wan, P.J.: Connected Dominating Set: Theory and Applications. Springer, Heidelberg (2013). https://doi.org/10.1007/978-1-4614-5242-3

23. Feige, U.: A threshold of ln n for approximating set cover. J. ACM **45**(4), 634–652 (1998). https://doi.org/10.1145/285055.285059. http://doi.acm.org/10.1145/285055.285059

24. Feige, U.: A threshold of ln n for approximating set cover. J. ACM **45**(4), 634–652 (1998). https://doi.org/10.1145/285055.285059

25. Feldkord, B., Markarian, C., Meyer Auf der Heide, F.: Price fluctuation in online leasing. In: Gao, X., Du, H., Han, M. (eds.) COCOA 2017. LNCS, vol. 10628, pp. 17–31. Springer, Cham (2017). https://doi.org/10.1007/978-3-319-71147-8_2

26. Garey, M.R., Johnson, D.S.: Computers and Intractability, A Guide to the Theory of NP-Completeness. W.H. Freeman and Company, New York (1979)

27. Garg, N., Vazirani, V.V., Yannakakis, M.: Primal-dual approximation algorithms for integral flow and multicut in trees, with applications to matching and set cover. In: Lingas, A., Karlsson, R., Carlsson, S. (eds.) ICALP 1993. LNCS, vol. 700, pp. 64–75. Springer, Heidelberg (1993). https://doi.org/10.1007/3-540-56939-1_62

28. Guha, S., Khuller, S.: Approximation algorithms for connected dominating sets. Algorithmica **20**(4), 374–387 (1998). https://doi.org/10.1007/PL00009201

29. Halldórsson, M.M.: Approximating k-set cover and complementary graph coloring. In: Cunningham, W.H., McCormick, S.T., Queyranne, M. (eds.) IPCO 1996. LNCS, vol. 1084, pp. 118–131. Springer, Heidelberg (1996). https://doi.org/10.1007/3-540-61310-2_10

30. Hamann, H., Markarian, C., Meyer auf der Heide, F., Wahby, M.: Pick, pack, & survive: charging robots in a modern warehouse based on online connected dominating sets. In: 9th International Conference on Fun with Algorithms, FUN 2018, La Maddalena, Italy, 13–15 June 2018, pp. 22:1–22:13 (2018). https://doi.org/10.4230/LIPIcs.FUN.2018.22

31. Hegeman, J., Ge, H., Gubin, M., Amit, A.: Sponsored advertisement ranking and pricing in a social networking system. US Patent 10,565,598, 18 February 2020
32. Hu, X., Ludwig, A., Richa, A.W., Schmid, S.: Competitive strategies for online cloud resource allocation with discounts: the 2-dimensional parking permit problem. In: 35th IEEE International Conference on Distributed Computing Systems, ICDCS 2015, Columbus, OH, USA, 29 June–2 July 2015, pp. 93–102 (2015). https://doi.org/10.1109/ICDCS.2015.18
33. Huang, L., Li, J., Shi, Q.: Approximation algorithms for the connected sensor cover problem. Theor. Comput. Sci. **809**, 563–574 (2020). https://doi.org/10.1016/j.tcs.2020.01.020. http://www.sciencedirect.com/science/article/pii/S0304397520300487
34. Johnson, D.S.: Approximation algorithms for combinatorial problems. J. Comput. Syst. Sci. **9**(3), 256–278 (1974). https://doi.org/10.1016/S0022-0000(74)80044-9. http://www.sciencedirect.com/science/article/pii/S0022000074800449
35. Kling, P., Meyer auf der Heide, F., Pietrzyk, P.: An algorithm for online facility leasing. In: Even, G., Halldórsson, M.M. (eds.) SIROCCO 2012. LNCS, vol. 7355, pp. 61–72. Springer, Heidelberg (2012). https://doi.org/10.1007/978-3-642-31104-8_6
36. Korman, S.: On the use of randomization in the online set cover problem. Master's thesis, Weizmann Institute of Science, Israel (2005)
37. Li, S., Markarian, C., Meyer auf der Heide, F.: Towards flexible demands in online leasing problems. Algorithmica **80**(5), 1556–1574 (2018). https://doi.org/10.1007/s00453-018-0420-y
38. de Lima, M.S., Felice, M.C.S., Lee, O.: On generalizations of the parking permit problem and network leasing problems. Electron. Notes Discret. Math. **62**, 225–230 (2017). https://doi.org/10.1016/j.endm.2017.10.039
39. Lovász, L.: On the ratio of optimal integral and fractional covers. Discret. Math. **13**(4), 383–390 (1975). https://doi.org/10.1016/0012-365X(75)90058-8. http://www.sciencedirect.com/science/article/pii/0012365X75900588
40. Lund, C., Yannakakis, M.: On the hardness of approximating minimization problems. J. ACM **41**(5), 960–981 (1994). https://doi.org/10.1145/185675.306789. http://doi.acm.org/10.1145/185675.306789
41. Markarian, C., Kassar, A.N.: Online deterministic algorithms for connected dominating set and set cover leasing problems. In: Proceedings of the 9th International Conference on Operations Research and Enterprise Systems, ICORES 2020, Valetta, Malta, 21–24 February 2020, pp. 121–128 (2020)
42. Markarian, C., Meyer auf der Heide, F.: Online algorithms for leasing vertex cover and leasing non-metric facility location. In: Parlier, G.H., Liberatore, F., Demange, M. (eds.) Proceedings of the 8th International Conference on Operations Research and Enterprise Systems, ICORES 2019, Prague, Czech Republic, 19–21 February 2019, pp. 315–321. SciTePress (2019). https://doi.org/10.5220/0007369503150321
43. Meyerson, A.: The parking permit problem. In: Proceedings of 46th Annual IEEE Symposium on Foundations of Computer Science (FOCS 2005), Pittsburgh, PA, USA, 23–25 October 2005, pp. 274–284 (2005). https://doi.org/10.1109/SFCS.2005.72
44. Munagala, K., Babu, S., Motwani, R., Widom, J.: The pipelined set cover problem. In: Eiter, T., Libkin, L. (eds.) ICDT 2005. LNCS, vol. 3363, pp. 83–98. Springer, Heidelberg (2004). https://doi.org/10.1007/978-3-540-30570-5_6
45. Nagarajan, C., Williamson, D.P.: Offline and online facility leasing. Discret. Optim. **10**(4), 361–370 (2013). https://doi.org/10.1016/j.disopt.2013.10.001
46. Yu, J., Wang, N., Wang, G., Yu, D.: Connected dominating sets in wireless ad hoc and sensor networks - A comprehensive survey. Comput. Commun. **36**(2), 121–134 (2013). https://doi.org/10.1016/j.comcom.2012.10.005

# Enabling Risk-Averse Dispatch Processes for Transportation Network Companies by Probabilistic Location Prediction

Keven Richly[✉], Rainer Schlosser, and Janos Brauer

Hasso Plattner Institute, University of Potsdam, Potsdam, Germany
keven.richly@hpi.de

**Abstract.** In the highly competitive ride-sharing market, optimized and cost-efficient dispatching strategies represent a crucial business advantage. A vulnerability of many state-of-the-art dispatch algorithms is the accuracy of the last observed location of available drivers that are used as input data. The current location of a driver is not exactly known since the observed locations can be outdated for several seconds and affected by noise. These inaccuracies affect dispatch decisions and cause critical delays. In this paper, we propose a prediction approach that provides a probability distribution for a driver's potential future locations based on patterns observed in past trajectories. We demonstrate the capabilities of our prediction results to (i) avoid critical delays, (ii) to estimate waiting times with higher confidence, and (iii) to enable risk-averse dispatching strategies. Furthermore, we developed a trajectory visualization tool to analyze the dispatch processes of a transportation network company and evaluated our approach on a real-world data set.

**Keywords:** Trajectory data · Location prediction algorithm · Peer-to-peer ride-sharing · Transport network companies · Risk-aware dispatching

## 1 Introduction

The demand for peer-to-peer ride-sharing services increased over the last years rapidly. Companies like Uber or Lyft offer a peer-to-peer ride-sharing service by connecting vehicle drivers with passengers to provide flexible and on-demand transportation [14]. Effective and efficient dispatching strategies that match drivers and passengers in real-time is a crucial component for a successful ride-sharing service [1]. The dispatching of requests focuses on reducing the overall travel time and waiting time of passengers, optimizing the utilization of available resources, and increasing the customer satisfaction [27]. Besides other criteria (e.g., customer status), spatio-temporal cost functions are an integral part of state-of-the-art dispatching algorithms. Based on the current position of all available drivers and the passenger, geographical distance or estimated time of arrival metrics are calculated [11]. Consequently, the quality of the decisions is strongly affected by the accuracy of the observed and transmitted locations of the drivers. For that reason, it is necessary to have exact location information or accurate location predictions.

© Springer Nature Switzerland AG 2022
G. H. Parlier et al. (Eds.): ICORES 2020/2021, CCIS 1623, pp. 21–42, 2022.
https://doi.org/10.1007/978-3-031-10725-2_2

## 1.1  Motivation

Based on a real-world data set of a globally operating transportation network company, we developed a visualization for trajectory data, which enabled the analysis and evaluation of dispatch processes. It empowers transportation network companies to identify limitations of dispatching policies and allows the comparison of different strategies. By inspecting the dispatching processes, reasons for unexpected critical delays can be investigated.

We identified that the inaccuracy and uncertainty of the exact locations used in the request dispatching process are one cause for delayed pick-up times and sub-optimal dispatch decisions. Surrounding urban effects cause signals to be noisy and lead to deviations of the recorded GPS location and the real one of a driver [26]. Additionally, the technical limitations of the GPS system and economic considerations constrain the emission of signals. The GPS locations of drivers are recorded and sent at specific intervals defined by the sampling rate to reduce bandwidth and storage costs. Furthermore, the entire dispatch process, including the acceptance confirmation by the driver, consumes several seconds, in which the driver is moving.

For that reason, it is necessary to develop new strategies to improve the accuracy of the applied spatio-temporal cost functions by optimized predictions of the exact drivers' locations. Furthermore, the dispatching algorithms should be enhanced and optimized by incorporating the risk factor for delays based on inaccurate positional information. By introducing risk-averse dispatching decisions, transportation network companies can optimize the request dispatching and avoid unnecessary detours of their drivers. Additionally, customer satisfaction can be increased by the communication of more accurate waiting times and the reduction of critical delays.

## 1.2  Contribution

This paper is an extended version of [21]. The main contributions of [21] are the following. We (i) implemented a trajectory visualization tool, which enables transportation network companies to analyze their dispatch processes and determine the causes of unexpected critical delays. We (ii) proposed a location prediction approach, which determines a distribution of potential future locations of drivers based on patterns observed in past trajectories. Compared to common dispatching algorithms that rely on outdated driver positions only, we are (iii) able to avoid critical delays by assigning drivers based on their estimated current potential position accounting for their individual driving behavior (speed, turn probabilities, etc.). We (iv) demonstrate that the prediction results allow to forecast potential waiting times with higher confidence.

Compared to [21], in this paper, we present extended studies and make the following contributions: First, as a short response time of the approach is crucial for practical applicability, we evaluate the prediction algorithm's runtime in greater detail. We studied how the algorithm's runtime is affected by a larger prediction time frame (of up to 30s). We find that although runtimes are increasing, which is due to the higher number of potential locations, the average response time remains sufficiently small (0.04 s) to be applied in practice. Second, we extend the dispatch framework by explicit risk considerations. We propose several risk-averse dispatching strategies, which seek to avoid the probability of critical delays. Instead of minimizing expected arrival times the different approaches minimize the probability of large waiting times, which helps to minimize the customers' cancellation rates.

This paper is organized as follows. In Sect. 2, we present related work. In Sect. 3, we describe the problem domain. Afterward, we present the developed application to analyze dispatch processes (Sect. 4). In Sect. 5, we present the limitations of dispatch decisions based on the last observed location of drivers. In Sect. 6, we describe our probabilistic location prediction approach. A numerical evaluation of the approach is presented in Sect. 7. Based on the results obtained, in Sect. 8, we propose different risk-aware dispatch strategies. Conclusions are given in the last section.

## 2   Related Work

In the following section, we review the literature form the related research fields route prediction and turning behavior prediction.

### 2.1   Route Prediction

Route prediction algorithms can be separated into long-term and short-term route prediction algorithms. Long-term route prediction approaches forecast drivers' entire route to their final destination, whereas short-term route prediction algorithms predict only a fraction of the remaining route a driver can drive within a provided prediction time. Various long-term route prediction algorithms use Hidden Markov Models (HMM) that model a driver's intended route as a sequence of hidden states since drivers' intentions can only be observed indirectly by the driven routes [10,23,28].

Simmons et al. [23] use an HMM that models the road segment, destination pairs as hidden states and the GPS data as observable states. While Simmons et al. [23] do not require a separate map-matching step, Ye et al. [28] require one, as their HMM models the driven road segment as observable states, while clusters of route serve as hidden states. Other approaches use clustering techniques to group similar trajectories into clusters so that the deviations of the current trajectories to past trajectories are more tolerated [3,10].

Lassoued et al. [10] hierarchically cluster trajectories via two different similarity metrics: same destination or route similarity metric. They define their route similarity metric as the fraction of shared road segment. Froehlich and Krumm [3] predict the intended route by using an elaborate route similarity function to compare the current route to a representative combination of routes of each cluster. The similarity metric depicts the distance differences between the GPS recordings of trajectory without pre-requiring a map-matching step. Further approaches use machine learning techniques, such as reinforcement learning [30], neural networks [15], and methods of social media analysis [28].

While long-term route prediction algorithms are helpful for the prediction of an entire route, their predictions are bound to previously observed routes. In our problem, however, the pick-up routes of individual drivers are rarely identical, as pick-up locations are not stationary, but various aspects can be used in short-term prediction.

Trasarti et al. [24] use clustering techniques to extract fractions the driver is expected to be able to drive within the provided prediction time. These approaches, however, still lack the support for new unseen routes.

Karimi et al. [8] predict the most probable short-term route by mining the driver's turning behavior at intersections and using the trajectories' underlying road network. They traverse the road network in depth-first fashion to find the maximum reachable locations from the driver's current location. They determine the traversal time of road segments by using the corresponding speed limits. This approach was extended by Jeung et al. [6] by mining the road segments' traversal time from trajectories. Both approaches require the trajectories to be map-matched, as the turn probabilities are calculated on the road segments level.

In contrast, Patterson et al. [17] avoid map-matching by using particle filters that incorporate the error of all random variables into one model. Additionally, dynamic short-term route algorithms exist, that reconstruct their models on-the-fly on data changes. These approaches acknowledge the dynamic nature of traffic and moving objects, whose environment changes aperiodically. Zhou et al. [29] continuously evict patterns from outdated observed trajectories so that the applied models only consider data from the most recent trajectories.

## 2.2 Turning Behavior Prediction

There are turning behavior predictions, which model drivers' turning behavior as a Markov process [6, 8, 9, 17, 31, 33]. These approaches are similar in the way they model the turning behavior at intersections as Markov chains, in which the states represent road segments, and drivers' decisions indicate their transitions at intersections. They differ, however, in the order of the Markov chain, i.e., the number of past road segments they consider.

While some consider only the last driven road segment to be an indicator for the next turn [6, 8, 17, 33], Krumm [9] proposes the usage of an $n^{th}$-order Markov chain, in which the next road segment is predicted by following the last $n$ driven road segments as states in the Markov chain. They evaluate that the more past road segments the prediction considers, the more accurate is the prediction of the turning behavior.

However, with the increasing order of the Markov chain, fewer sequences of driven road segments are observed, as the Markov state space increases exponentially. Also, they experimented with inferring if the result's accuracy is sensitive to context information, such as time of day or day of the week. However, they did not find such sensitivity, as the fraction of matched road segment sequences of the given context was small due to the training dataset's size.

Ziebart et al. [31] model the turning behavior of drivers via a Markov decision process whose cost weight of actions are learned via inverse reinforcement learning using context- and road-specific features. Further approaches analyze the speed and acceleration profiles of drivers to predict the turning behavior at an upcoming intersection.

Liebner et al. [12] cluster speeding profiles using k-means to predict a driver's turning behavior at a single intersection. Phillips et al. [18] and Zyner et al. [32] use short-term memory neural networks to predict the turning behavior.

# 3   Background

In this section, we define all relevant information entities that are part of the problem domain and necessary to understand the visualization concepts as well as the proposed algorithm to avoid risky dispatches.

A road network is a directed multigraph that represents real-world traffic infrastructure of a specified area along with the corresponding metadata [2]. In the graph, each node represents an intersection between at least two road segments, which are represented by edges. These road network maps are created and maintained by humans or automatically updated by trajectory-based algorithms [4]. The meta-information includes, for example, the length and speed limit of a road segment as well as the exact geographic locations for all intersections and road segments [2].

**Definition 1.** *Road Network: A road network is a multigraph $R$ represented by a 4-tuple $R = (I, E, \Sigma_I, \Sigma_E)$. $I$ is a set of nodes representing intersections. $\Sigma_I$ and $\Sigma_E$ contain the node and edge labels, respectively. $E \subseteq V \times V \times \Sigma_E$ is the set of edges encoding road segments between intersections. The node labels $\Sigma_I$ are composed of an intersection's GPS location, whereas the edge labels $\Sigma_E$ consist of a road segment's geographic extent, length, and speed limit.*

**Definition 2.** *Road Segment: A road segment $r$ is a directed edge that is confined by a source $r.source$ and target $r.target$ intersection. It is associated with a list of intermediate GPS points describing the segment's geography. Each road segment contains a length and a speed limit. A set of connecting road segment composes a road.*

In this work, a trajectory is a chronologically ordered sequence of map-matched and timestamped observed locations of a driver, which represents a continuous driving session. For that reason, we use a segmentation algorithm to split the raw positional data of a single moving object into separate trajectories. The start and end of driving sessions are defined by events like changes of the occupancy state or inactive time intervals of drivers.

**Definition 3.** *Trajectory: A trajectory $T_d^{t_s, t_e}$ is a chronologically ordered sequence of map-matched and timestamped observed locations of a driver $d$ in a given time interval $[t_s, t_e]$.*

**Definition 4.** *Ping: A ping $p_d^t$ depicts a map-matched observed location of a driver $d$ at time $t$. The state $p_d^t$ is given by a 3-tuple $(l, s, t)$, denoting that the driver $d$ is located at location $l$ with the occupancy state $s$ at time $t$. The location $l$ consists of the tuple $(x, y)$ representing the map-matched GPS coordinates with longitude and latitude.*

As mentioned in the previous section, the accuracy of GPS locations is affected by various factors (e.g., noise) [26]. For that reason, it is possible that the observed locations of a driver are off-road. Therefore, we use common map-matching algorithms to match the locations to a reference road network. For each observed location a map-matched location on a road segment is determined based on the trajectory of a driver.

**Fig. 1.** A screenshot of the application, displaying the trajectory of the driver (orange and blue dots), the fastest route (green line), and the predicted next locations via purple circles, cf. [21]. (Color figure online)

## 4    Visualization of Dispatch Processes

With the capabilities to display the trajectory data, our application enables transportation network companies (i) to analyze dispatch decisions, (ii) to evaluate and compare different dispatching algorithms, (iii) to determine the effect and accuracy of location prediction algorithms, and (iv) to label spatio-temporal data for comprehensive investigations or as foundation for machine learning approaches. Through the detailed analysis of past dispatches, it is possible to identify reasons for late pick-ups and determine characteristics of scenarios, in which the risk for a delay exists. Additionally, it provides the opportunity to identify general problems of dispatch strategies and to examine the behavior in edge cases.

### 4.1    Analyzing Dispatch Decisions

An overview of bookings enables transportation network companies to navigate through various dispatch processes to identify problematic dispatches efficiently (e.g., significant delays). The bookings can be filtered and sorted by different criteria (e.g., delay, manually assigned labels) to select specific dispatches to be analyzed in more detail (see Fig. 1).

In the analysis view, the system visualizes the spatio-temporal data associated with the dispatch decision on a map. As shown in Fig. 1, the trajectory of the assigned driver (colored dots), the position of the pick-up location (black marker), and the shortest route to the pick-up location (green line) are displayed. Additionally, the corresponding information (e.g., estimated time of arrival determined by the transportation network company and the Open Source Routing Machine (OSRM)[1]) are shown. The trajectory of the driver is divided into orange and blue dots, which represent the associated pings. The color change indicates the timestamp at which the driver acknowledged the transportation request, and the trip was assigned. Consequently, the orange points represent the free-time trajectory of the driver. In periods without passengers or passenger requests, the drivers drive freely around intending to get in an excellent position to be selected by the dispatch algorithm for the next booking request.

The blue dots of the trajectory represent the route of the driver after the assignment of the trip. Here, the driver has a particular target location and tries to reach the pick-up location on the shortest path. By comparing this trajectory with the shortest path determined by OSRM, the user has a good indicator of problematic dispatches. As displayed in the example (see Fig. 1), the delay of the driver was caused by an initial detour. Furthermore, we can analyze the circumstances around the assignment of the trip and determine potential reasons for the detour (e.g., inaccurate positional information or a driver's position on a road segment, which makes it impossible for him to drive the shortest route). In Sect. 5, we discuss these issues in more detail.

To evaluate different prediction algorithms as well as our probabilistic approach (described in Sect. 6), the application visualizes the predicted locations along with the determined probabilities. The locations are displayed directly on the map to allow the user to compare the predicted positions (purple circles) with the last observed location and the trajectory of the driver after the dispatch process.

## 4.2 Determining the Estimated Fastest Pickup Routes

The application illustrates the fastest route between the last ping of the dispatched driver's free-time trajectory and pick-up location as a solid line. We use the OSRM, a tool of the OpenStreetMap community, to calculate a driver's fastest pick-up route. In contrast to routing services used by deployed dispatching algorithms of transportation network companies, the routing functionality of OSRM is not traffic-adjusted. Instead, it estimates the cost of a road segment, i.e., its traversal time, as its length divided by its speed limit. The traversal speed estimation via the speed limit is a significant simplification, as the scenario that a driver traverses the road network without any traffic and with traversal speed indicated by the speed limit is very unlikely.

However, this constraint is acceptable, as even if we use the same traffic-adjusted routing service as the deployed dispatching algorithm, the calculated pick-up route and its traversal time may differ from the route the dispatching algorithm has retrieved at the time of the dispatch from the same service. The reason is that routing services, such as Google Maps, incorporate traffic in real-time to keep estimates accurate and hence, the suggested fastest route for the same pair of GPS coordinates changes continuously with the underlying traffic. The fastest pick-up route that we retrieve from OSRM is not guaranteed to be identical to the pick-up route that was used by the dispatching algorithm.

---

[1] http://project-osrm.org.

Hence, the estimated traversal time of the fastest pick-up route and the estimated traversal time calculated by the dispatching algorithm of the transportation network company are not comparable to each other.

## 5   Improving Dispatch Decisions by Probabilistic Location Prediction Algorithms

As already mentioned, it is necessary to provide exact location information of all available drivers to communicate accurate pick-up times to passengers and to efficiently assign passengers to drivers. The assignment of available drivers to requesting passengers in the context of transportation network companies is a dynamic vehicle routing problem or dial a ride problem.

The vehicle routing problem is characterized as dynamic, if requests are received and updated concurrently with the determination of routes, see Psaraftis et al. [19]. In the setup of transportation network companies, new passenger requests have to be continuously assigned to available drivers considering further information, such as the current traffic situation or the availability of drivers, which are unknown in advance. For that reason, companies are applying different policies typically intending to optimize specific objective functions (e.g., to minimize the overall waiting time of passengers or route costs) [20].

Correspondingly, the applied policy to select a driver from a set of available drivers is based on a cost function (e.g., minimum costs, minimum distance, minimum travel time, maximum number of passengers). Most of these functions use the location of the passengers and the location of the available drivers as inputs. A common example is the nearest vehicle dispatch, which assigns the passenger request to the driver with the shortest travel time to the pick-up location [7]. Based on the locations, the travel time is determined by using services that offer traffic-adjusted routing services (e.g., Google Maps).

For that reason, accurate calculations require precise and up-to-date location information about all available drivers. However, there are different factors like noise or technical limitations of GPS system [26].

Additionally, the given sampling rate, data transfer problems, and the time consumed by the entire process affects the accuracy of the spatio-temporal information. Consequently, the actual position of a driver at the time of the order assignment can deviate significantly from the last observed location, which is currently used as input to calculate the estimated travel time or distance.

### 5.1   Limitations of Status - Quo Dispatch Decisions

To demonstrate the limitations of dispatch decisions based on the last observed location, we use the dispatching example depicted in Fig. 2 to exemplify the implications of the inaccuracy and uncertainty of a driver's current location at the time of dispatch. The example shows a dispatching scenario on a highway, where the upper-right user pin represents the passenger's pick-up location and the car pins represent a single driver's GPS locations. While the dotted marker represents the driver's last recorded location (which the dispatching algorithm uses), the solid markers represent the driver's possible locations at the time of dispatch. The driver's last recorded position in the example is affected by noise so that the recorded location resides between the two highway lanes. Depending on its implementation, the dispatching algorithm may now assume that the driver is on the right lane, however, if the driver's correct location is $A$, the actual travel time can be much higher than its estimated counterpart, as turns on highways are impossible and the next exit may be far away.

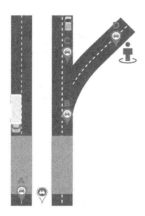

**Fig. 2.** An example highlighting the implications of the driver's current location's inaccuracy and uncertainty. The dotted location marker between the two highway lanes depicts the last recorded location. The other markers indicate a driver's possible current locations on the two roads, cf. [21].

Even when on the right side of the street, the driver's location at the time of dispatch relative to the necessary highway exit is unknown: the driver may have or may not have taken the exit (location $D$ and $C$), or the driver may not have reached the exit (location $B$). The actual travel time varies significantly with locations $B - D$, as missed exists on highways are costly in terms of time. Consequently, there is a high risk of delay. Additionally, the driver's last recorded location may be older than indicated by the sampling rate or urban effects, such as tunnels, prevent the emission of GPS signals. Also, the entire process of assigning a driver and the acknowledgment of the drive takes several seconds, where the position of the driver is continuously changing.

As shown by the example, an inaccuracy and uncertainty of the drivers' locations at the time of dispatch can significantly influence the determined value of the cost function (e.g., travel time). Therefore, the dispatching algorithm has to decide based on incorrect information, for which reason it may not assign the optimal driver to a requesting passenger and also the driver could arrive delayed at the pick-up location. For that reason, we introduce the concept of Detoured Dispatches and Risky Dispatches, see below.

**Definition 5.** *Detoured Dispatch: A dispatch is classified as a detoured dispatch if the assigned driver's arrival at the pick-up location is delayed due to an initial detour of the driver.*

**Definition 6.** *Risky Dispatch: A dispatch is said to be risky if the dispatched driver's arrival at the pick-up location is likely to be delayed due to uncertainty about the current position of a driver, which may lead to an initial detour or a sub-optimal route.*

After the selection of a driver, the exact current position is also necessary to calculate the estimated waiting time, which is communicated to the customer. The waiting time has to be accurate as the cancellation rate strongly increases with the displayed waiting time. High cancellation rates reflect unsatisfied passengers leading to a drop in passenger retention rate, as the industry of ride-hailing is characterized by fierce competition. Ultimately, high cancellation rates reduce the revenue of a transportation network company. The communicated waiting time has to be accurate, i.e., the actual travel time cannot be much longer than the calculated travel time. Otherwise, the passenger has to wait longer than initially communicated, leading to an increase in the cancellation rate. We observed that passengers do not tolerate delays, as more than 50% of all delay-related cancellations happen within the first two minutes of a delay.

To evaluate the share of delays caused by detoured dispatches, we analyzed a sample of 500 dispatch decisions with our application manually. The dispatch processes were randomly selected from a real-world dataset of a transportation network company, which includes the bookings and the spatio-temporal data of Dubai, spanning from November 2018 to February 2019. Further, we limited the analysis to dispatch processes where the driver arrived at the pick-up location between one and five minutes delayed. We classified a dispatch as detoured if the driver performed an initial detour after the confirmation of the trip and returned to the determined fastest route afterward. Based on the random sample, we identified that in about 20% of the delayed arrivals, the driver performed an initial detour.

## 5.2 Probabilistic Location Predictions and Implications for Dispatch Decisions

An example of how probabilistic location predication can influence the dispatch decisions is shown in Fig. 3. The black marker represents the pick-up location and the blue, green, and orange markers the last observed map-matched location of three available drivers. A traditional dispatching algorithm that uses a specific cost function (e.g., shortest distance or shortest travel time) would assign the booking request to the blue driver based on the last observed locations. By analyzing the predicted potential positions of the drivers, we can see that the blue and green drivers are likely to move away from the location of the passenger. In contrast, the orange driver is directly driving in the direction of the passenger. For that reason, it is highly likely that the orange driver would be the best option for the algorithm.

In this example, we demonstrate that by including the driving behavior and direction of drivers, the result of the dispatch algorithm can change. Additionally, we can immediately detect whether the estimated time of arrival of a certain driver (e.g., the blue driver in Fig. 4) would be too optimistic and detours and, in turn, critical delays are likely.

In the second example (see Fig. 4), we demonstrate the impact of the probabilities calculated based on observed patterns in past drives. The size of the dots represents the probability of the corresponding location. The larger a dot is, the higher is the probability of the location. Similar to the first example (see Fig. 4), the blue driver has the

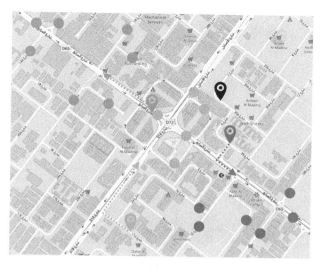

**Fig. 3.** Predicting potential current locations of candidate drivers to be assigned to a waiting customer (black marker): Example of three different drivers (green, blue, orange marker). The dots represent predicted potential next locations of each driver based on their driving behavior, cf. [21]. (Color figure online)

shortest distance and seemingly the shortest travel time to the pick-up location. But the big dot in the left-bottom corner indicates that there is a high chance that the blue driver misses the exit. For that reason, it may be preferable to assign the trip to another driver.

The green driver has a higher probability of being on the shortest route to the pick-up location, but also there is a not negligible probability that the driver stays on the highway and needs to perform a costly detour to reach the location of the passenger.

Based on the last observed location, the orange driver has the longest distance to the pick-up location, but the predicted probabilities show that she is highly likely driving the direction of the pick-up location. Consequently, to assign the order to the orange driver is potentially not the optimal decision, but the one with the lower risk of delays.

Our proposed approach enables transportation network companies to apply dispatching strategies that take risk considerations into account. Whether to optimize expected arrival times, worst-case scenarios, or other risk-aware criteria can be strategically determined by the companies. Our approach, however, is a key for such risk-aware dispatching strategies.

## 6   Probabilistic Location Prediction Algorithm

To minimize detoured dispatches and enable risk-aware decisions, we propose a model to predict probabilities of future driver positions based on patterns observed in past trajectories. We suggest the algorithm to be used to predict the possible locations of dispatching candidates at the time of assignment of the trip. The dispatching algorithm calculates the estimated travel time from a combination of travel times considering the set of possible locations. By mining historic drives and predicting possible locations

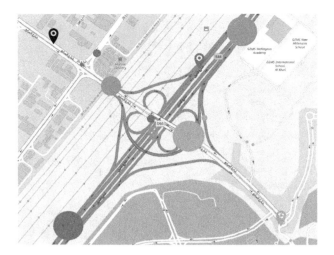

**Fig. 4.** Improving dispatch decisions using probability distributions for the current locations of potential drivers: Comparing the likelihood of a driver to reach the customer (black marker) without critical delays. Example of three different drivers (green, blue, orange marker). The dots represent the predicted next locations of each driver (the larger the dot is, the higher is the probability of the location), cf. [21]. (Color figure online)

allows for a more precise estimation of pick-up times leading to shorter waits, in spite of the inherent uncertainty and inaccuracy of a driver's current position.

The goal of this approach is to observe repeating driving patterns from all drivers that can be generalized so that we can apply them to forecast upcoming driving behaviors. The generalization requires the analysis of past driving behavior that is representative of future behavior. As we forecast a driver's next locations around the time of dispatch, we constrain the analysis' dataset to free-time trajectories. In free-time trajectories, drivers are generally not influenced by external factors and thus can drive freely around.

At the time of dispatch, drivers are unaware of a request until it is communicated to them, which is after the dispatch process. Consequently, at the time of dispatch drivers drive freely around, and hence, their decisions are similar to the ones taken before in past free-time trajectories. The analysis of trajectories also allows us to extract information on the dynamic characteristics of the road network, such as traffic. Traffic affects drivers' traversal times on road segments and hence we need to incorporate this into the location prediction to ensure accuracy. Traffic repeats itself [25], we can use historical traffic patterns to forecast future traversal times on road segments consequently.

Overall, the prediction algorithm consists of the five parts (i) data preprocessing, (ii) map matching, (iii) road segment candidates determination, (iv) route probability calculation, and (v) location prediction. Next, we describe the different steps.

### 6.1 Data Preprocessing

During the data preprocessing, we segment the trajectories in sub-trajectories that represent distinct driving sessions and extract the sub-trajectories with the occupancy state free. Afterward, we map-match the observed locations to retrieve their actual location on a road segment in the road network. Based on the map-matched pings, we interpolate the route between subsequent pings if their road segments are discontiguous.

Depending on the occupancy state, the driving behavior of a driver changes significantly. If the driver is transporting passengers or is on the way to pick-up passengers, she is driving the shortest route based on the current position, the destination, and the current traffic situation. These routes are often suggested by routing services.

In contrast, drivers with the occupancy state *free* are freely driving around with the goal of getting incoming bookings. Their routes are depending on personal experience and individual preferences as well as external circumstances. For that reason, we have to distinguish trajectories based on the occupancy state for our use case.

**Definition 7.** *Occupancy State of Trajectory: The occupancy state of a trajectory $T_d^{t_s,t_e}$ is defined by the state of all pings of the trajectory. For that reason, all pings of a trajectory must have the same occupancy state. We distinguish between the two states available and occupied.*

We define a route as an ordered sequence of connected road segments, which are determined by the trajectory and defines a semantic compression of the trajectory consequently. Multiple consecutive pings on a road segment are combined. Additionally, if the resulting road segments are not connected, the corresponding road segments to connect the segments by the shortest path are added to the route.

**Definition 8.** *Route: A route $R_d^{t_s,t_e}$ of a driver d is a sequence of connected road segments, visited by driver d in the time interval $[t_s, t_e]$ ordered by the time of traversal.*

A booking represents a transportation request from a passenger. During dispatch, a potential driver is assigned to the booking. After the driver confirms the booking, her occupancy state changes from available to occupied. Accordingly, the state changes to available after the driver finished a booking.

## 6.2 Map Matching

The accuracy of the observed GPS locations can be affected by various factors [26]. Noise, limitations of the GPS technology, and measurement errors can cause a discrepancy between the observed and the exact location of a driver. As displayed in Fig. 2, these inaccuracies can have the consequence that an observed location is off-road or assigned to an incorrect road segment. For that reason, a map-matching step is necessary to ensure that each observed location is assigned to a road segment on the one hand and, on the other hand, to reduce inaccuracies and filter measurement errors [13]. To match the locations to a reference road network, we use the established map-matching library *Barefoot*[2].

Additionally, we apply filters to remove physically implausible sequences of map-matched locations. Newson and Krumm [16], also suggest filter and cleansing approaches for outliers (e.g., traversal speed and maximum acceleration thresholds).

---

[2] https://github.com/bmwcarit/barefoot.

### 6.3  Road Segment Candidates

After the map matching step, we determine a set of road segment candidates. Instead of considering just all road segments in the neighborhood, this approach restricts the set of candidates that include all potential reachable ones to a possible small number of road segments. Due to this step, the number of candidates is reduced for subsequent parts of the prediction algorithm, which in turn allows for faster predictions.

To determine the relevant road segments on which the driver is estimated to be after the defined prediction frame, we partially analyze the road network. Based on the last observed location of an available driver, we calculate all possible paths of the driver by summing up the traversal times of the road segments until the prediction frame is exceeded. By definition, the algorithm expects drivers to reach the last road segment of a path, and we add the last road segment to the list of candidates consequently.

Accordingly, each road segment has an associated cost, which depicts its traversal time. The traversal time is defined as the time a driver needs to traverse a road segment completely. There are different approaches to specify the traversal time (e.g., speed limits and average speed of a driver). We use a method that determines the traversal time based on the traversal speed from past trajectories.

The mined traversal speed is the average speed of all drivers on the road segment in a given time frame. As the traffic volume changes in specific areas over the day (e.g., rush hour), the daytime has an impact on the traversal speed of a road segment. The selection of a small time frame enables the consideration of detailed traffic patterns but also strongly limits the number of observed drivers in the time frame. For that reason, we selected a time frame of one hour, which allows us to distinguish between different traffic situations and delivers a sufficient number of observations for the majority of road segments.

Due to the fact that we consider all pings of a driver on a specific road segment, the mined traversal speed implicitly includes various traffic effects like traffic light phases or traffic jams. In cases in which we have no or only a minimal number of observations, we use the current speed of the driver if all road segments in the path have the same speed limit. The last fallback option is to use the speed limits.

### 6.4  Route Probability Calculation

To determine the probability of a route, we analyze drivers' turning behavior at intersections to assess drivers' probability to take a specific turn. These turning patterns allow the computation of the overall probability of different potential routes. By definition, intersections connect at least two road segments. Consequently, to model drivers' turning behavior at intersections, we can count the co-occurrences of road segment pairs and calculated the corresponding probabilities [8, 9, 34].

We model the turn probabilities by a Markov chain of $n^{th}$-order, as a driver's behavior at intersections can be represented by a sequence of events, in which the probability of each event, i.e., the decision at the current intersection, depends only on the state attained in the previous event, i.e., the decision at the previous intersection. In the model, the road segments serve as states and the turn probabilities as transition probabilities between the states.

The selection of the Markov chain order is a tradeoff. Markov chains of a higher order allow us to represent the behavior of drivers better to drive around a specific area. This behavior is typical for drivers of transportation network companies, as particular regions are more profitable compared to others [22]. Using $n^{th}$-order Markov chains, the number of routes, i.e., states of the Markov chain, that do not have many observations increases, since the number of possible states increases exponentially with the order. A road network contains a large number of road segments. Therefore, we may observe some road segments of the road network infrequently. States with few observations are inexpressive for drivers' turn behavior. To avoid inexpressive states, we filter states whose support is below a threshold.

As already mentioned, traffic situations and the demand vary for specific regions depending on the daytime [22]. For that reason, we distinguish between different contexts (e.g., rush hour) to increase the accuracy of the turn probability prediction.

## 6.5 Location Extrapolation on Road Segment Candidates

At the last step, we extrapolate a driver's specific location on the determined road segments, as the estimated time to the passenger can vary based on the particular location on a road segment. During the short-term route prediction, we calculate a set of road segments candidates that a driver is expected to reach within the prediction frame $f$. We determine for each candidate road segment a driver's required traversal time $t_{path} \in \mathbb{R}_0^+$ to reach it. As the remaining time $t_{remaining}$ of each candidate road segment, i.e., $t_{remaining} = f - t_{path}$, is not large enough to traverse it completely, we expected the driver to be located on it. Given each candidates remaining and traversal time, we estimate the drivers detailed position via the fraction of the road segment the driver is expected to have traversed within the prediction frame.

# 7 Evaluation

In this section, we evaluate the accuracy of our location prediction algorithm. We perform out-of-sample four-fold cross-validation for all experiments and report the average score over all four runs. Further, we evaluate the algorithm's runtime for prediction frames of different lengths.

## 7.1 Experimental Setup

To evaluate our approach, we used a real-world trajectory dataset of a renowned transportation network company. The dataset includes observed locations of drivers and booking information in the city of Dubai, spanning from November 2018 to February 2019. Compared to publicly available datasets, it has a high sampling rate of 5 s. For the period, we have over 400 million observed locations.

Based on the time span between two observed locations, we segment the trajectory data of a driver in sub-trajectories representing continuous driving session. We classified the sub-trajectories based on the occupancy state and get $\approx 1.5$ million free-time trajectories. The OpenStreetMap road network of Dubai has about 139K road segments with average length of 115 m.

## 7.2 Accuracy of the Prediction Algorithm

We evaluate the overall quality of the next location prediction algorithm. We use 1 000 pings located on the same road segment, to predict the road segments their associated drivers could be on after the prediction frame, i.e., the road segment candidates, along with their respective probabilities. The drivers' correct road segment after the prediction frame serves as the ground truth.

We compare the discrete probability distribution of these predicted road segments with the discrete relative frequency distribution of the drivers' correct road segments of the ground truth. We model the turning behavior via $2^{nd}$-order Markov chains. For the experiments, we set the prediction frame to 5, 10, and 20 s and evaluate the algorithm's performance for a representative example.

In Fig. 5, we illustrate the results of our location prediction algorithm for drivers that are currently on a frequented road segment. The training dataset for the algorithm includes 21 751 traversal speed observations and 9 224 turn observations for the respective road segments. The predicted probability density over the set of road segment candidates is similar to the distribution of the relative frequencies of the individual road segments of the ground truth. The average absolute difference between the probability of a predicted road segment and its relative frequency in the ground truth for the prediction frames are small: 0.012 (5 s), 0.033 (10 s), and 0.075 (20 s). The result verifies the accuracy of our approach.

For a prediction frame of 5 s, the predicted probability deviates on average by 1.2% from the actual relative frequency. The difference proves that the location prediction algorithm is accurate for frequently observed road segments. As the prediction frame increases, the difference of the predicted probabilities of the road segments to their actual relative frequencies increases. The reason for this is that with increasing prediction frame, the impact of the estimated traversal speeds' inaccuracies increases. The imprecision of the estimation may be caused by temporary traffic conditions that the mined traversal speed estimations do not capture in full detail.

We conducted further location predictions for different examples. Naturally, we found that the results depend on the specific setting considered (road segment, time, individual driving behavior, etc.). However, overall, we obtained similar accuracy results as in the shown example, see Fig. 6. Further, we observed that the most critical factor is the amount of data associated with a specific setting.

Moreover, we evaluated if the turning behavior at intersections changes with the time of the day (e.g., rush hour). For that reason, we construct one Markov chain that models the turning behavior of drivers during rush hour and one during the evening hours. We select these hours so that both Markov chains cover the same number of observations.

Further, to assess if the context-specific modeling boosts prediction accuracy, we assess if Markov chains of the same context have more similar turn probabilities than Markov chains of different contexts, considered as *Both*, cf. Fig. 6. We measure the similarity via the average absolute difference of turn probabilities of the same Markov state. We constrain the comparison to intersections with at least 50 observations for each context. The restriction results in 2 485 intersections, for which we compare the turn probabilities.

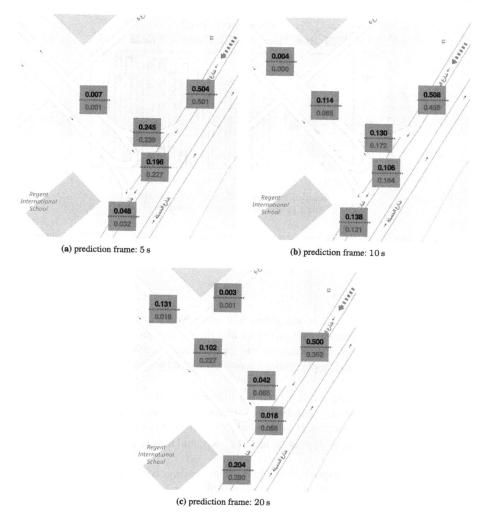

(a) prediction frame: 5 s

(b) prediction frame: 10 s

(c) prediction frame: 20 s

**Fig. 5.** Results of the next location prediction algorithm on a representative road segment. We run the experiment on 1 000 out-of-sample pings that share the same road segment indicated by the dashed green arrow. The upper value in the box denotes the relative frequency of drivers that are on the respective road segment after the prediction frame. In contrast, the value below depicts the probability we predict for drivers to be on that road segment after the prediction frame, cf. [21]. (Color figure online)

The results, see Fig. 6, show that the estimated turn probabilities are accurate for different contexts, i.e., rush hour and evening. For both contexts, around 80% of all Markov states' turn probabilities have at most an average absolute difference of 0.05. In contrast, the Markov chains differ across contexts more significantly. Only around 65% of the Markov states' turn probabilities have at most an average absolute difference of 0.05. In contrast to Krumm [9], our results demonstrate that including context information can improve the accuracy of turn probabilities.

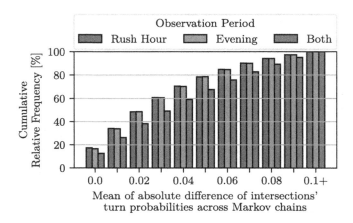

**Fig. 6.** Sensitivity to context: The histogram shows the cumulative distribution of the mean absolute differences of intersections' turn probabilities of different times of the day, cf. [21].

### 7.3   Runtime Evaluation

Dispatching algorithms have to decide in real-time. Note, with increasing passenger waiting time, the cancellation rate increases and the discrepancy between drivers' last recorded locations used for dispatching and their current location increases with execution time. To verify the applicability of our approach, we evaluate the algorithm's runtime.

The prediction algorithm consists of the five parts (i) data preprocessing, (ii) map matching, (iii) road segment candidates determination, (iv) turn probability calculation, and (v) location prediction. Most importantly, the final prediction of a driver's probabilistic location takes not more than 30 milliseconds, and hence, is applicable in real-life settings.

As the algorithm's components (i) data preprocessing, (ii) map matching, and (iv) turn probability calculation can be processed offline and updated from time to time, the critical components are (iii) road segment candidates determination and (v) location prediction, cf. Sect. 6. While the final location prediction is simple and fast, the determination road segment candidates the highest runtime complexity. Hence, in the following evaluation, we focussed on component (iii).

We investigated the effect of an increasing prediction frame on the runtime of the prediction algorithm. We executed the algorithm with a prediction frame of 5, 10, and 30 s. Naturally, with a larger prediction frame the algorithm has to evaluate more potential road segments and the runtime increases. However, we observe that the execution times are still sufficiently low to be applied in practice. The average runtime for a prediction frame of 5 and 10 s, respectively, is 0.027 s and 0.032 s. For a prediction frame of 30 s, the algorithm requires on average 0.040 s.

## 8   Risk-Averse Dispatch Strategies

Based on the probabilistic prediction results derived, in the previous section, we propose different risk-aware dispatch strategies. When choosing a driver the goal is to not only minimize the (expected) arrival time but also account for the risk of critical delays. By $p_k^{(d)}$ we denote the probability that a driver $d$'s current location is $k$ (where the set of potential locations $K^{(d)}$ is derived as described in Sect. 6); by $t_k^{(d)}$ we denote the (estimated) time to reach the customer from location $k$. A driver $d$'s random arrival time is denoted by $T^{(d)}$. Next, we define five different dispatch assignment strategies, see $A$ - $E$.

### 8.1   Best Expected Arrival Time

Select the driver $d$ with the smallest expected arrival time $E(T^{(d)})$ $=$ $\sum_{k \in K^{(d)}} p_k^{(d)} \cdot t_k^{(d)}$, i.e.,

$$\min_d E(T^{(d)})$$

The approach optimizes the mean arrival time based on potential current locations instead of the outdated last known location as used in many status quo strategies, see Sect. 5.1.

### 8.2   Worst Case Optimization

Select the driver $d$ with the smallest worst case arrival time:

$$\min_d \left\{ \max_{k \in K^{(d)}} t_k^{(d)} \right\}$$

The approach seeks to keep potential delays as small as possible. The approach allows to determine an upper bound for the arrival time.

### 8.3   Mean-Variance Optimization

Select the driver $d$ with the best balance of mean arrival time and associated variance penalized by a chosen factor $\alpha$, e.g., $\alpha := 0.1$:

$$\min_d \left\{ E(T^{(d)}) + \alpha \cdot \sum_{k \in K^{(d)}} p_k^{(d)} \cdot \left( t_k^{(d)} - E(T^{(d)}) \right)^2 \right\}$$

The criteria combines the risk-neutral approach $A$ with an incentive to choose drivers with low deviations of potential arrival times according to their uncertain current location, which in turn, yields more plannable arrival times.

### 8.4  Best Expected Utility

Select the driver $d$ with the best expected utility (i.e., smallest expected penalty) using a suitably chosen convex penalty function $u$, e.g., $u(x) := x^2$:

$$\min_d \left\{ \sum_{k \in K^{(d)}} p_k^{(d)} \cdot u(t_k^{(d)}) \right\}$$

The criteria looks for small arrival times while penalizing delays in a progressive way (as intended). Similar techniques are used in [5] for risk-sensitive path selections under uncertain travel costs.

### 8.5  Probability Constraints and Quantile-Based Criteria

Select the driver $d$ with the best arrival time that can be realized with at least probability $z$ (smallest $z$-quantile):

$$\min_d \left\{ q \,\Big|\, P\left(T^{(d)} \leq q\right) \geq z \right\}$$

The criteria allows to formulate performance measures that are easy to interpret and to communicate, e.g., with $z = 90\%$ probability a driver will arrive within time $q$.

For all proposed approaches the computation time is not an issue as each objective can be easily evaluated numerically for a couple of potential drivers. The proposed approaches allow to optimize different risk preferences, which may resemble the strategic goals of a company. In this context, also further criteria can be used (e.g., (conditional) value at risk, etc.). Note, the key for all of those approaches is our probabilistic location prediction. Naturally, the suitability and effectiveness of the proposed risk-aware dispatch strategies have to be evaluated in practice. This will be addressed in future work.

## 9  Conclusion

We studied limitations of applied dispatching strategies using an application to visualize the trajectory data of drivers in the period of dispatch processes. Our tool supports transportation network companies to derive a deeper understanding of reasons for unexpected critical delays caused by inefficient dispatch decisions. In particular, we identified inaccurate and outdated positional information as one aspect for the late arrivals of drivers at the pick-up location. These inaccuracies are produced by various circumstances (e.g., noise, technical limitations).

Further, we address this problem by proposing a location prediction approach that provides a probability distribution for a driver's future locations based on patterns observed in past trajectories. More specifically, we are able to quantify with which probability a driver has moved in which direction since the last ping. Furthermore, our approach allows to also account for personalized and time-dependent driving characteristics.

We demonstrate the applicability and the accuracy of our prediction approach using numerical experiments based on a real-life dataset. Moreover, we verify that the algorithm's runtime is sufficiently fast to be applied in practice.

In order to further improve traditional dispatch strategies, we propose different concepts to minimize the probability of large waiting times. This way, the risk of critical delays can be addressed, and thus, helps to minimize the customers' cancellation rates.

# References

1. Agatz, N., Erera, A., Savelsbergh, M., Wang, X.: Optimization for dynamic ride-sharing: a review. Eur. J. Oper. Res. **223**(2), 295–303 (2012)
2. Ben Ticha, H., Absi, N., Feillet, D., Quilliot, A.: Vehicle routing problems with road-network information: state of the art. Networks **72**(3), 393–406 (2018)
3. Froehlich, J., Krumm, J.: Route prediction from trip observations. In: SAE Technical Paper (2008)
4. He, S., et al.: Roadrunner: improving the precision of road network inference from GPS trajectories. In: Proceedings of the 26th ACM SIGSPATIAL International Conference on Advances in Geographic Information Systems, pp. 3–12. ACM (2018)
5. Hu, J., Yang, B., Guo, C., Jensen, C.S.: Risk-aware path selection with time-varying, uncertain travel costs-a time series approach. VLDB J. **27**, 179–200 (2018)
6. Jeung, H., Yiu, M.L., Zhou, X., Jensen, C.S.: Path prediction and predictive range querying in road network databases. VLDB J. **19**(4), 585–602 (2010)
7. Jung, J., Jayakrishnan, R., Park, J.Y.: Design and modeling of real-time shared-taxi dispatch algorithms. In: Proceedings of the Transportation Research Board's 92nd Annual Meeting (2013)
8. Karimi, H.A., Liu, X.: A predictive location model for location-based services. In: Proceedings of the 11th International Symposium on Advances in Geographic Information Systems, New Orleans, Louisiana, USA, pp. 126–133. ACM (2003)
9. Krumm, J.: A Markov model for driver turn prediction (2016)
10. Lassoued, Y., Monteil, J., Gu, Y., Russo, G., Shorten, R., Mevissen, M.: A Hidden Markov model for route and destination prediction. In: 20th IEEE International Conference on Intelligent Transportation Systems, ITSC 2017, Yokohama, Japan, pp. 1–6. IEEE (2017)
11. Liao, Z.: Real-time taxi dispatching using global positioning systems. Commun. ACM **46**(5), 81 (2003)
12. Liebner, M., Baumann, M., Klanner, F., Stiller, C.: Driver intent inference at urban intersections using the intelligent driver model. In: Proceedings of the 2012 Intelligent Vehicles Symposium, IV 2012, Alcal de Henares, Madrid, Spain, pp. 1162–1167. IEEE (2012)
13. Lou, Y., Zhang, C., Zheng, Y., Xie, X., Wang, W., Huang, Y.: Map-matching for low-sampling-rate GPS trajectories. In: Proceedings of the 17th ACM SIGSPATIAL International Conference on Advances in Geographic Information Systems, pp. 352–361 (2009)
14. Masoud, N., Jayakrishnan, R.: A real-time algorithm to solve the peer-to-peer ride-matching problem in a flexible ridesharing system. Transp. Res. Part B: Methodol. **106**, 218–236 (2017). https://doi.org/10.1016/j.trb.2017.10.006
15. Mikluščák, T., Gregor, M., Janota, A.: Using neural networks for route and destination prediction in intelligent transport systems. In: Mikulski, J. (ed.) TST 2012. CCIS, vol. 329, pp. 380–387. Springer, Heidelberg (2012). https://doi.org/10.1007/978-3-642-34050-5_43
16. Newson, P., Krumm, J.: Hidden Markov map matching through noise and sparseness. In: Proceedings of the 17th ACM SIGSPATIAL International Conference on Advances in Geographic Information Systems, pp. 336–343. ACM (2009)

17. Patterson, D.J., Liao, L., Fox, D., Kautz, H.: Inferring high-level behavior from low-level sensors. In: Dey, A.K., Schmidt, A., McCarthy, J.F. (eds.) UbiComp 2003. LNCS, vol. 2864, pp. 73–89. Springer, Heidelberg (2003). https://doi.org/10.1007/978-3-540-39653-6_6

18. Phillips, D.J., Wheeler, T.A., Kochenderfer, M.J.: Generalizable intention prediction of human drivers at intersections. In: Proceedings of the 2017 Intelligent Vehicles Symposium, Los Angeles, California, USA, pp. 1665–1670, June 2017

19. Psaraftis, H.N.: Dynamic vehicle routing: status and prospects. Ann. Oper. Res. **61**(1), 143–164 (1995)

20. Psaraftis, H.N., Wen, M., Kontovas, C.A.: Dynamic vehicle routing problems: three decades and counting. Networks **67**(1), 3–31 (2016)

21. Richly, K., Brauer, J., Schlosser, R.: Predicting location probabilities of drivers to improve dispatch decisions of transportation network companies based on trajectory data. In: Proceedings of the 9th International Conference on Operations Research and Enterprise Systems - ICORES. vol. 1, pp. 47–58. SciTePress (2020). https://doi.org/10.5220/0008911100470058

22. Richly, K., Teusner, R.: Where is the money made? An interactive visualization of profitable areas in New York city. In: The 2nd EAI International Conference on IoT in Urban Space (Urb-IoT) (2016)

23. Simmons, R.G., Browning, B., Zhang, Y., Sadekar, V.: Learning to predict driver route and destination intent. In: Intelligent Transportation Systems Conference, ITSC 2006, pp. 127–132. IEEE, September 2006

24. Trasarti, R., Guidotti, R., Monreale, A., Giannotti, F.: MyWay: location prediction via mobility profiling. Inf. Syst. **64**, 350–367 (2017)

25. Treiber, M., Kesting, A.: Traffic Flow Dynamics. Traffic Flow Dynamics: Data, Models and Simulation (2013)

26. Wang, Y., Zhu, Y., He, Z., Yue, Y., Li, Q.: Challenges and opportunities in exploiting large-scale GPS probe data. HP Laboratories, Technical Report HPL-2011-109 21 (2011)

27. Xu, Z., et al.: Large-scale order dispatch in on-demand ride-hailing platforms: a learning and planning approach. In: Proceedings of the 24th ACM SIGKDD International Conference on Knowledge Discovery & Data Mining, pp. 905–913. ACM (2018)

28. Ye, N., Wang, Z.Q., Malekian, R., Lin, Q., Wang, R.C.: A method for driving route predictions based on hidden Markov model. Math. Probl. Eng. **2015**, 1–12 (2015)

29. Zhou, J., Tung, A.K., Wu, W., Ng, W.S.: A "semi-lazy" approach to probabilistic path prediction. In: Proceedings of the 19th International Conference on Knowledge Discovery and Data Mining, Chicago, Illinois, USA, p. 748 (2013)

30. Ziebart, B.D., Maas, A.L., Bagnell, J.A., Dey, A.K.: Maximum entropy inverse reinforcement learning. In: Proceedings of the 23rd Conference on Artificial Intelligence, Chicago, Illinois, USA, pp. 1433–1438, July 2008

31. Ziebart, B.D., Maas, A.L., Dey, A.K., Bagnell, J.A.: Navigate like a cabbie: probabilistic reasoning from observed context-aware behavior. In: Proceedings of the 10th International Conference on Ubiquitous Computing, Seoul, Korea, pp. 322–331, September 2008

32. Zyner, A., Worrall, S., Ward, J.R., Nebot, E.M.: Long short term memory for driver intent prediction. In: Intelligent Vehicles Symposium, IV 2017, Los Angeles, California, USA, pp. 1484–1489. IEEE, June 2017

33. Liu, X., Karimi, H.A.: Location awareness through trajectory prediction. Comput. Environ. Urban Syst. **30**(6), 741–756 (2006)

34. Wang, X., et al.: Building efficient probability transition matrix using machine learning from big data for personalized route prediction. Procedia Comput. Sci. **53**, 284–291 (2015). Elsevier

# Solving the Robust Vehicle Routing Problem Using the Worst Feasible Scenario Strategy

Zuzana Borčinová[(✉)] and Štefan Peško

Department of Mathematical Methods and Operations Research, Faculty of Management Science and Informatics, University of Žilina, Univerzitná 8215/1, Žilina, Slovakia
zuzana.borcinova@fri.uniza.sk

**Abstract.** This paper deals with the Robust Capacitated Vehicle Routing Problem where customer demands are uncertain and belong to predetermined uncertainty set. The general approach of robust optimization is to optimize the worst instance that might arise due to data uncertainty. However, if the worst instance is infeasible, then there is no solution that can meet all demand realizations. Therefore we propose a strategy, which seeks a solution that satisfies as many demands as possible with respect to vehicle fleet capacity. The computational experiments examined our proposed strategy to indicate its performance in terms of the extra cost and unmet demands.

**Keywords:** Capacitated vehicle routing problem · Uncertain demands · Robust optimization · Worst-case scenario

## 1 Introduction

The Capacitated Vehicle Routing Problem (CVRP) is one of the combinatorial optimization problems which aims to find a set of minimum total cost routes for a fleet of capacitated vehicles based at one depot, to serve a set of customers under the following constraints:

(1) each route begins and ends at the depot,
(2) each customer is visited exactly once,
(3) the total demand of each route does not exceed the capacity of the vehicle [15].

The first mathematical formulation and algorithm for the solution of the CVRP was proposed by Dantzig and Ramser [7] and five years later, Clarke and Wright [6] proposed the first heuristic for this problem. Till to date many solution methods for the CVRP have been published. General surveys can be found in [14, 27]. The CVRP belongs to the category of NP-hard problems that can be exactly solved only for small instances of the problem. Therefore, researchers have concentrated on developing heuristic algorithms to solve this problem, for example [8, 13].

Supported by the research grants VEGA 1/0342/18 "Optimal dimensioning of service systems", APVV-19-0441 "Allocation of limited resources to public service systems with conflicting quality criteria" and VEGA 1/0776/20 "Vehicle routing and scheduling in uncertain conditions".

© Springer Nature Switzerland AG 2022
G. H. Parlier et al. (Eds.): ICORES 2020/2021, CCIS 1623, pp. 43–55, 2022.
https://doi.org/10.1007/978-3-031-10725-2_3

Contrary to the deterministic CVRP, which assumes that the problem parameters (e.g. the customer demands, travelling costs, service times, etc.) are deterministic and known, the robust CVRP (RVRP) considers the parameters affected by uncertainty with unknown probability distribution. In robust optimization methodology introduced by Ben-Tal and Nemirovski [1], uncertainty is modeled as a bounded set $U$ which contains all possible continuous or discrete values referred as scenarios. The objective of the RVRP is to obtain a robust solution, which is feasible for all scenarios in $U$.

There are some works dealing with the RVRP with different (combined or separated) uncertain parameters. For example, Sungur et al. [25], Moghaddam et al. [18], Gounaris et al. [9,10] and Pessoa et al. [20] study the RVRP with uncertain demands. Toklu et al. [26], Han et al. [12] and Solano-Charris et al. [23] apply robust optimization to the RVRP with uncertain travel costs. Lee et al. [16] and Sun et al. [24] consider RVRP in which travel times and also demands are uncertain.

This paper is an extended version of study published in [5], which deals with the RVRP with uncertain customer demands and the distribution of them is unknown. Our research is inspired by a work of Sungur et al. [25], who proposed the first solution procedure for the RVRP with demand uncertainty. The authors used a robust version of the CVRP formulation with Miller-Tucker-Zemlin constraints based on specific uncertainty to determine vehicle routes that minimize transportation costs while satisfying all possible demand realizations. Their model may be perceived as a worst-case instance of the deterministic CVRP in which the nominal demand parameter is replaced by a modified one from the uncertainty set, thus solving the RVRP is not more difficult than solving a single deterministic CVRP. But, depending on the nature of the scenarios, some RVRPs may become infeasible problems. Therefore, we are interested in extending of this method to infeasible instances. We extend our previous work by effective iterative solution method and additional experiments.

The structure of this paper is organized as follows. Section 2 is devoted to the RVRP formulation with uncertain demands which is derived from our two-index CVRP formulation [4]. In Sect. 3, our robust solution approach is presented. In the next session, the computational experiments are described and results are reported and compared with adopted method of Sungur et al. Finally in Sect. 5, conclusions are drawn.

## 2    Formulation of the RVRP with Uncertain Demands

The RVRP with demand uncertainty can be defined by a complete directed graph $G = (V, E, c)$ with $V = \{0, 1, 2, \ldots, n\}$ as the set of nodes and $E = \{(i, j) : i, j \in V, i \neq j\}$ as the set of arcs, where node 0 represents the depot for a fleet of $p$ vehicles with the same capacity $Q$ and remaining $n$ nodes represent geographically dispersed customers.

The positive travel cost $c_{ij}$ is associated with each arc $(i, j) \in E$. The cost matrix is symmetric, i.e. $c_{ij} = c_{ji}$ for all $i, j \in V, i \neq j$ and satisfies the triangular inequality, $c_{ij} + c_{jl} \geq c_{il}$ for all $i, j, l \in V$.

The uncertain demands associated with each customer $j \in V - \{0\}$ are modeled as a set of discrete scenarios. The scenarios are constructed as values around an expected demand $d_j^0$ with the maximum perturbation $\varepsilon$, where $\varepsilon$ is a nonnegative constant. Thus demand $d_j^k$ of customer $j$ in scenario $k$ is positive value, $d_j^0 - \varepsilon d_j^0 \leq d_j^k \leq d_j^0 + \varepsilon d_j^0$.

Two-index decision variables $x_{ij}$ are used as binary variables equal to 1 if arc $(i, j)$ belongs to the optimal solution and 0 otherwise. For all pairs of nodes $i, j \in V - \{0\}, i \neq j$ we calculate the savings $s_{ij}$ for joining the cycles $0 \rightarrow i \rightarrow 0$ and $0 \rightarrow j \rightarrow 0$ using arc $(i, j)$ as in Clarke and Wright's saving method [6], i.e.

$$s_{ij} = c_{i0} + c_{0j} - c_{ij}. \tag{1}$$

Then, instead of minimizing the total cost, we maximize the total saving [4].

**Theorem 1.** *Maximizing the total travel saving* $\sum\limits_{\substack{i=1}}^{n} \sum\limits_{\substack{j=1 \\ i\neq j}}^{n} s_{ij}\, x_{ij}$ *is equivalent to mini-*

*mizing the total travel cost* $\sum\limits_{\substack{i=0}}^{n} \sum\limits_{\substack{j=0 \\ i\neq j}}^{n} c_{ij}\, x_{ij}.$

*Proof.* The total travel cost can be expressed in the form:

$$\sum\limits_{\substack{i=0}}^{n} \sum\limits_{\substack{j=0 \\ i\neq j}}^{n} c_{ij}\, x_{ij} = \sum\limits_{i=1}^{n} c_{i0}\, x_{i0} + \sum\limits_{j=1}^{n} c_{0j}\, x_{0j} + \sum\limits_{\substack{i=1}}^{n} \sum\limits_{\substack{j=1 \\ i\neq j}}^{n} c_{ij}\, x_{ij}.$$

According to (1) $c_{ij} = c_{i0} + c_{0j} - s_{ij}$ and we obtain:

$$\sum\limits_{\substack{i=0}}^{n} \sum\limits_{\substack{j=0 \\ i\neq j}}^{n} c_{ij}\, x_{ij} = \sum\limits_{i=1}^{n} c_{i0}\, x_{i0} + \sum\limits_{j=1}^{n} c_{0j}\, x_{0j} + \sum\limits_{\substack{i=1}}^{n} \sum\limits_{\substack{j=1 \\ i\neq j}}^{n} c_{i0}\, x_{ij} + \sum\limits_{\substack{i=1}}^{n} \sum\limits_{\substack{j=1 \\ i\neq j}}^{n} c_{0j}\, x_{ij} - \sum\limits_{\substack{i=1}}^{n} \sum\limits_{\substack{j=1 \\ i\neq j}}^{n} s_{ij}\, x_{ij}.$$

By a simple manipulation we get:

$$\sum\limits_{\substack{i=0}}^{n} \sum\limits_{\substack{j=0 \\ i\neq j}}^{n} c_{ij}\, x_{ij} = \sum\limits_{\substack{i=1}}^{n} \sum\limits_{\substack{j=0 \\ i\neq j}}^{n} c_{i0}\, x_{ij} + \sum\limits_{\substack{i=0}}^{n} \sum\limits_{\substack{j=1 \\ i\neq j}}^{n} c_{0j}\, x_{ij} - \sum\limits_{\substack{i=1}}^{n} \sum\limits_{\substack{j=1 \\ i\neq j}}^{n} s_{ij}\, x_{ij},$$

which can be rewritten as:

$$\sum\limits_{\substack{i=0}}^{n} \sum\limits_{\substack{j=0 \\ i\neq j}}^{n} c_{ij}\, x_{ij} = \sum\limits_{i=1}^{n} c_{i0} \sum\limits_{\substack{j=0 \\ i\neq j}}^{n} x_{ij} + \sum\limits_{j=1}^{n} c_{0j} \sum\limits_{\substack{i=0 \\ i\neq j}}^{n} x_{ij} - \sum\limits_{\substack{i=1}}^{n} \sum\limits_{\substack{j=1 \\ i\neq j}}^{n} s_{ij}\, x_{ij}.$$

Since each customer is served by exactly one vehicle, it means that

$$\sum\limits_{\substack{i=0 \\ i\neq j}}^{n} x_{ij} = \sum\limits_{\substack{i=0 \\ i\neq j}}^{n} x_{ji} = 1$$

for all $j \in V - \{0\}$. It implies that:

$$\sum\limits_{\substack{i=0}}^{n} \sum\limits_{\substack{j=0 \\ i\neq j}}^{n} c_{ij}\, x_{ij} = \sum\limits_{i=1}^{n} c_{i0} + \sum\limits_{j=1}^{n} c_{0j} - \sum\limits_{\substack{i=1}}^{n} \sum\limits_{\substack{j=1 \\ i\neq j}}^{n} s_{ij}\, x_{ij}$$

and so

$$\sum_{\substack{i=0 \\ i\neq j}}^{n}\sum_{\substack{j=0 \\ i\neq j}}^{n}c_{ij}\,x_{ij} + \sum_{\substack{i=1 \\ i\neq j}}^{n}\sum_{\substack{j=1 \\ i\neq j}}^{n}s_{ij}\,x_{ij} = \sum_{i=1}^{n}c_{i0} + \sum_{j=1}^{n}c_{0j} = \text{const.}$$

Thus we have shown that the sum of total travel cost and total travel saving is constant, therefore maximizing the total travel saving minimizes the total travel cost.    □

To ensure the continuity of the route and to eliminate sub-tours, we define an auxiliary continuous variable $y_j^k$, which value is (in the case of collecting of the goods) the vehicle load after visiting customer $j$ in scenario $k$. To simplify model formulation, we replace each feasible route $0 \to v_1 \to v_2 \to \cdots \to v_t \to 0$ by a path from node $0$ to node $v_t$, i.e. $0 \to v_1 \to v_2 \to \cdots \to v_t$.

The RVRP with uncertain demands seeks the optimal solution that satisfies all demand realizations. Let $U_d$ denotes a set which contains all scenario vectors, $U_d = \{d^k, k = 1, \ldots, q\}$, where $q$ is a number of scenarios. Then, according to Bertsimas and Sim [2], the robust optimization model of the problem can be stated as

*RVRP 1:*

$$\max \quad \sum_{i=1}^{n}\sum_{\substack{j=1 \\ i\neq j}}^{n}s_{ij}\,x_{ij}, \tag{2}$$

subject to

$$\sum_{j=1}^{n}x_{0j} = p, \tag{3}$$

$$\sum_{i=1}^{n}x_{i0} = 0, \tag{4}$$

$$\sum_{\substack{i=0 \\ i\neq j}}^{n}x_{ij} = 1, \quad \forall j \in V - \{0\}, \tag{5}$$

$$\sum_{\substack{j=1 \\ i\neq j}}^{n}x_{ij} \leq 1, \quad \forall i \in V - \{0\}, \tag{6}$$

$$y_i^k + d_j^k\,x_{ij} - Q\,(1 - x_{ij}) \leq y_j^k, \tag{7}$$
$$\forall k \in \{1, \ldots, q\}, i, j \in V - \{0\}, i \neq j,$$

$$d_j^k \leq y_j^k \leq Q, \quad \forall k \in \{1, \ldots, q\}, j \in V - \{0\}, \tag{8}$$

$$x_{ij} \in \{0, 1\}, \quad \forall i, j \in V, i \neq j. \tag{9}$$

In this formulation (see [5]) the objective function (2) maximizes the total travel saving. The constraints (3), (4), (5) and (6) are the indegree and outdegree constraints for depot and customers. Constraints (7) are the route continuity and sub-tour elimination constraints and the constraints given in (8) are capacity bounding constraints which restrict the upper and lower bounds of $y_j^k$. Finally, (9) are the obligatory constraints.

The general approach of robust optimization is to optimize the worst-case value over all data uncertainty [25]. Let $d_j^w$ denotes the demand of customer $j$ in the worst scenario. Let us replace the constraints (7) and (8) with the constraints

$$y_i^w + d_j^w x_{ij} - Q\left(1 - x_{ij}\right) \le y_j^w,$$

$$\forall i, j \in V - \{0\}, i \ne j, \tag{10}$$

$$d_j^w \le y_j^w \le Q, \quad \forall j \in V - \{0\}. \tag{11}$$

The continuous variable $y_j^w$ has the same meaning as in *RVRP 1*, i.e. $y_j^k$ is the load of vehicle when demand value of customer $j$ attain their worst case. Thus, similar to Sungur et al. [25], we can solve the RVRP as an instance of the CVRP. We refer to the resulting model as *RVRP 2*.

## 3   Determining the Worst-Case Scenario

The key step in this approach is to identify the worst-case scenario and subsequently use its values $d_j^w$ in the constraints (10) and (11). We suggest two strategies for determining the worst-case scenario: the maximum demand scenario and the worst feasible demand scenario [5].

### 3.1   The Maximum Demand Scenario

In the first strategy (*Strategy 1*) we consider that all customer demands take their maximal values in each scenario [19], i.e. the worst-case scenario values are $d_j^w = d_j^{max} = \max\{\max_k d_j^k, d_j^0\}$ for $j \in V - \{0\}$, $k \in \{1, \dots, q\}$, where $q$ is a number of scenarios. This strategy solution is immune to demand variation under each scenario. However, the CVRP with the maximum demand scenario may be infeasible if vehicles have not sufficient capacity. In this case, robust solution can not be found.

### 3.2   The Worst Feasible Demand Scenario

The main idea of this strategy (*Strategy 2*) is to serve as many demands as possible, i.e. we find such scenario with values $d_j^w \in \{d_j^k, k = 0, \dots, q\}$, which maximizes the sum of satisfied demands with respect to the fleet size and vehicle capacities.

We introduce the binary decision variables $t_{ijk}$, which indicate whether vehicle $i$ serves demand $d_j^k$ or not. The mathematical model that determines feasible scenario with maximal sum of satisfied demands can be described as follows:

$$\max \quad \sum_{i=1}^{p} \sum_{j=1}^{n} \sum_{k=0}^{q} t_{ijk}\, d_j^k, \tag{12}$$

subject to

$$\sum_{j=1}^{n}\sum_{k=0}^{q}t_{ijk} \geq 1, \quad \forall i \in \{1,\dots,p\}, \tag{13}$$

$$\sum_{i=1}^{p}\sum_{k=0}^{q}t_{ijk} = 1, \quad \forall j \in \{1,\dots,n\}, \tag{14}$$

$$\sum_{j=1}^{n}\sum_{k=0}^{q}t_{ijk}\, d_j^k \leq Q, \quad \forall i \in \{1,\dots,p\}, \tag{15}$$

$$t_{ijk} \in \{0,1\}, \tag{16}$$
$$\forall i \in \{1,\dots,p\}, j \in \{1,\dots,n\}, k \in \{0,\dots,q\}.$$

The objective function (12) maximizes the sum of the demand values selected from a set of scenarios to be served. The constraints (13) mean that each vehicle must serve at least one customer. The constraints (14) impose that every customer is visited once by one vehicle and exactly one of its demand value is selected to be served. The constraints (15) ensure that the total load does not exceed the capacity of any vehicle. Finally, the constraints (16) define the binary decision variables. Obviously, there is no guarantee that all the demands for each customer will be met under this strategy.

## 4    Computational Experiments

In order to assess the efficiency and quality of the proposed robust solution approach, we designed computational experiments and analyzed the robust solutions obtained with alternative strategies in terms of their demand and cost performances. The mathematical models were coded in Python 3.7 [21] and solved by the solver Gurobi 8.1 [11]. A computer with an Intel i7 8 cores, 3 GHz processor and 32 GB of RAM was used to perform the computational experiments.

### 4.1    Test Instances

For our computational experiments we used standard instances publicly available at http://www.coin-or.org/SYMPHONY/branchandcut/VRP/data/. Because the instances were originally designed for deterministic CVRP, it was necessary to modify them to include demand uncertainty. The customer demands specified in the benchmark were taken to be their nominal values $d_j^0$. For each deterministic CVRP benchmark, we construct four classes of uncertainty sets of 5 scenarios within the allowed perturbation $\varepsilon \in \{0.05, 0.1, 0.15, 0.2\}$.

### 4.2    Exact Solution Method

Since the demand values in the worst-case scenario are not smaller than nominal values, the *RVRP 2* is more tight (capacity constrained) than corresponding CVRP. Thus solving the *RVRP 2* is more difficult than solving the deterministic version. To facilitate computing we use our Mixed Integer Programming (MIP) iterative method for the CVRP [3] based on the concept of $k$-opt:

**Definition 1.** *A solution is said to be $k$-optimal (or simply $k$-opt) if it is impossible to obtain a better solution by replacing any $k$ of its edges by any other set of $k$ edges [17].*

From this definition it is obvious that any $k$-opt tour is also $k'$-opt for $1 \leq k' \leq k$. It is also easy to show that a tour containing $m$ edges is optimal if and only if it is $m$-opt. Given a feasible solution, our solution method repeatedly performs exchanges of $k$ edges that reduce the cost of the current solution, until a solution is reached for which no exchange leads to an improvement. Then obtained solution is $k$-opt. The value of $k$ changes dynamically according to the following algorithm:

**Algorithm**

1: Apply a fast heuristic method to find an initial (feasible) CVRP solution $S_0$. Set $S = S_0$.
2: Create a list $E$ of arcs which belong to solution $S$ and are not incident with the depot. Let $m = |E|$.
3: Set values $t = 1, m_1 = m - \delta$ and $m_2 = m - 1$, where $\delta$ is a predetermined integer value $1 < \delta < m$.
4: Find an optimal solution $S_t$, so that the set $E_t$ of arcs which belong to solution $S_t$ contains minimally $m_1$ and maximally $m_2$ of arcs from $E$, i.e. we add to *RVRP 2* the constraints:

$$\sum_{(i,j) \in E} x_{ij} = z, \tag{17}$$

$$m_1 \leq z \leq m_2, \tag{18}$$

where integer variable $z$ determines how many arcs from $E$ are retained in $E_t$.
5: If the solution $S_t$ is better than the solution $S$, then set $S = S_t$ and continue to step 2, else set values $t = t + 1, m_1 = m_1 - \delta$ a $m_2 - \delta$ and go to step 4.
6: If $m_1 < 0$, set $m_1 = 0$. If $m_2 \geq 0$, go to step 4, else Stop.

The algorithm terminates when $0 \leq z \leq 1$ (or equivalently $m - 1 \leq k \leq m$) and no better solution was found. It means that final solution is optimal. Of course, the algorithm can be stopped after a predetermined computational time or number of iterations without improvement. In that case, obtained solution is $k$-optimal, $k < m$.

In our experiments we computed the initial feasible solution using parallel version of Clarke and Wright's saving (CWS) method [6]. This heuristic starts from a solution where each customer is served by a different route. At each iteration the CWS algorithm tries to merge two routes, one with arc $(i, 0)$ and second with arc $(0, j)$, maximizing the saving $s_{ij}$ (1), under the condition that the merged route is feasible.

The instances with $n < 50$ customers were solved to optimality, for other ones the solution method was terminated after 3 iterations without improvement.

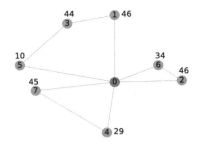

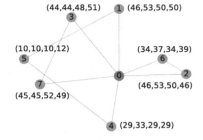

**Fig. 1.** CVRP: Optimal deterministic solution.        **Fig. 2.** RVRP: Robust solution.

### 4.3   Performance Measures

To evaluate the performance of the proposed strategies, we use the performance measures presented in [25] including the relative extra cost and unmet demands.

**Cost Performance.** Let $z_r$ and $z_d$ be the cost of robust and deterministic solutions respectively. The ratio

$$\zeta = \frac{z_r - z_d}{z_d} \tag{19}$$

quantifies the relative extra cost of the robust solution versus the deterministic optimal one. It is clear, that smaller $\zeta$ means the better cost performance.

**Demand Performance.** The unmet demand is the sum of demands in each route that exceeds the vehicle capacity. Let $g_d$ and $g_r$ represent the maximum unmet demand that can occur when using the deterministic and robust solution respectively and $D$ is the total nominal demand, i.e.

$$D = \sum_{i=1}^{n} d_i^0. \tag{20}$$

The ratio

$$\gamma = \frac{g_d - g_r}{D} \tag{21}$$

reflects the relative reduction of unmet demand in the robust solution compared to deterministic optimal one when the maximum demand scenario occurs. Obviously, larger $\gamma$ indicate better demand performance. We notice that for every solution found by *Strategy 1 $g_r = 0$.*

We explain these terms by the following illustrative example [5].

*Example 1.* Figure 1 shows a CVRP optimal deterministic solution with $n = 7$ customers and $p = 3$ vehicles of the capacity $Q = 100$, where the customer demands

$$d = (46, 46, 44, 29, 10, 34, 45)$$

are displayed next to the nodes. The cost of this solution is $z_d = 227$.

Figure 2 illustrates a robust solution of the same problem with $m = 4$ discrete scenarios of uncertain demands, which were generated randomly by perturbation $\varepsilon = 0.20$:

$$d^0 = (46, 46, 44, 29, 10, 34, 45),$$

$$d^1 = (53, 53, 44, 33, 10, 37, 45),$$

$$d^2 = (50, 50, 48, 29, 10, 34, 52),$$

$$d^3 = (50, 46, 51, 29, 12, 39, 49).$$

The route demands in particular scenarios are:

Route 1, 5, 4: 85, 96, 89, 91
Route 3, 7: 89, 89, 100, 100
Route 2, 6: 80, 90, 84, 85

It is evident, that the total demand of any route in each scenario does not exceed vehicle capacity. The cost of robust solution $z_r = 291$, i.e. the relative extra cost $\zeta = 0.282$.

Since *Strategy 1* failed to solve this problem, abovementioned robust solution was found by *Strategy 2*, whereby the maximum feasible scenario is

$$d^w = (53, 53, 51, 33, 12, 39, 49).$$

To evaluate the demand performance, we calculate the total demands for each route in the case that all customer demands take their maximum value

$$d^{max} = (53, 53, 51, 33, 12, 39, 52).$$

The sums of demands in the optimal solution routes are 116, 85 and 92, i.e. there is $g_d = 16$ unsatisfied demands. In the robust solution, the sums of demands in the routes are 98, 103 and 92, therefore $g_r = 3$ and the relative reduction of unmet demand $\gamma = 0.051$.

## 4.4   Numerical Results

The results of both proposed strategies are summarized in the Table 1.

In this table, the first column represents the name of instances. The name of each instance allows to determine its characteristics, because it has a format X-nA-kB-eC, where A is the number of nodes, B represents the number of vehicles and C indicates perturbation. For example an instance A-n32-k5-e15 has 32 nodes, 5 vehicles and was derived from the instance A-n32-k5 by demand generation with $\varepsilon = 0.15$.

The columns *Cost performance* and *Demand performance* show the two performance measures $\zeta$ and $\gamma$ respectively, as explained above. An indicator "*in*" denotes an infeasible instance.

As we can observe from Table 1 the strategy, which optimizes the maximum demand scenario (*Strategy 1*) leads to infeasible solutions in some cases, while strategy based on the maximum feasible demand scenario approach (*Strategy 2*) has an appropriate solution in all cases. For all other cases both strategies have achieved the same results,

**Table 1.** The comparison of two strategies.

| Instance | $D$ | $z_d$ | $g_d$ | Strategy 1 | | | | Strategy 2 | | | |
|---|---|---|---|---|---|---|---|---|---|---|---|
| | | | | $z_r$ | $g_r$ | Cost perform. | Demand perform. | $z_r$ | $g_r$ | Cost perform. | Demand perform. |
| A-n32-k5-e5 | 410 | 784 | 0 | 784 | 0 | 0.000 | 0.000 | 784 | 0 | 0.000 | 0.000 |
| A-n32-k5-e10 | 410 | 784 | 4 | 825 | 0 | 0.052 | 0.010 | 825 | 0 | 0.052 | 0.010 |
| A-n32-k5-e15 | 410 | 784 | 16 | 808 | 0 | 0.031 | 0.039 | 808 | 0 | 0.031 | 0.039 |
| A-n32-k5-e20 | 410 | 784 | 30 | 850 | 0 | 0.084 | 0.073 | 850 | 0 | 0.084 | 0.073 |
| A-n34-k5-e5 | 460 | 778 | 0 | 778 | 0 | 0.000 | 0.000 | 778 | 0 | 0.000 | 0.000 |
| A-n34-k5-e10 | 460 | 778 | 0 | 778 | 0 | 0.000 | 0.000 | 778 | 0 | 0.000 | 0.000 |
| A-n34-k5-e15 | 460 | 778 | 2 | 790 | 0 | 0.015 | 0.004 | 790 | 0 | 0.015 | 0.004 |
| A-n34-k5-e20 | 460 | 778 | 13 | in | in | in | in | 811 | 11 | 0.042 | 0.004 |
| A-n44-k6-e5 | 570 | 937 | 0 | 937 | 0 | 0.000 | 0.000 | 937 | 0 | 0.000 | 0.000 |
| A-n44-k6-e10 | 570 | 937 | 2 | 939 | 0 | 0.002 | 0.004 | 939 | 0 | 0.002 | 0.004 |
| A-n44-k6-e15 | 570 | 937 | 11 | 983 | 0 | 0.049 | 0.019 | 983 | 0 | 0.049 | 0.019 |
| A-n44-k6-e20 | 570 | 937 | 20 | in | in | in | in | 978 | 15 | 0.044 | 0.009 |
| A-n53-k7-e5 | 664 | 1010 | 6 | 1033 | 0 | 0.023 | 0.009 | 1033 | 0 | 0.023 | 0.009 |
| A-n53-k7-e10 | 664 | 1010 | 5 | 1033 | 0 | 0.023 | 0.008 | 1033 | 0 | 0.023 | 0.008 |
| A-n53-k7-e15 | 664 | 1010 | 26 | in | in | in | in | 1059 | 3 | 0.049 | 0.035 |
| A-n53-k7-e20 | 664 | 1010 | 54 | in | in | in | in | 1056 | 37 | 0.046 | 0.026 |
| A-n60-k9-e5 | 829 | 1354 | 0 | 1354 | 0 | 0.000 | 0.000 | 1354 | 0 | 0.000 | 0.000 |
| A-n60-k9-e10 | 829 | 1354 | 4 | 1360 | 0 | 0.004 | 0.005 | 1360 | 0 | 0.004 | 0.005 |
| A-n60-k9-e15 | 829 | 1354 | 9 | 1373 | 0 | 0.014 | 0.011 | 1373 | 0 | 0.014 | 0.011 |
| A-n60-k9-e20 | 829 | 1354 | 25 | in | in | in | in | 1373 | 18 | 0.014 | 0.008 |

because the maximum feasible demand scenario is equal to the maximum demand scenario. It means, that *Strategy 2* is an extension of *Strategy 1* for problems with infeasible maximum demand scenario.

The deterministic solutions of some instances are robust, since $g_d = 0$, so in these cases $\zeta$ and $\gamma$ are zero. For other instances, the robust solution has a larger cost than the optimal solution for the nominal demand, but on the other hand this optimal solution can not meet more customer's demand when maximum demand scenario occurs. In practice, the higher solution cost means the lower current profit, while the lower unmet demand implies the higher customer satisfaction and thus the higher future profit. The trade-off between decreasing the unmet demand and increasing the solution cost is illustrated in Fig. 3.

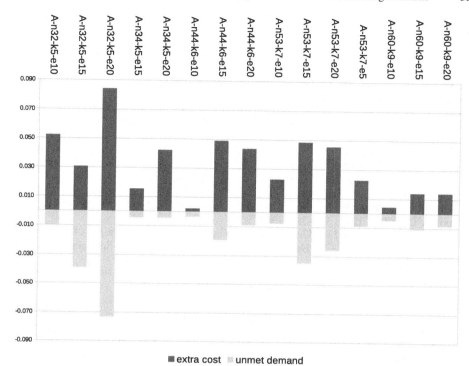

**Fig. 3.** Relative reduction of unmet demand versus relative extra cost.

## 5   Conclusions

In this paper robust optimization was used to solve the Capacitated Vehicle Routing Problem with demand uncertainty. The main contribution of our work is in developing a new strategy to cope with uncertain demands. At first, we present the mathematical model *RVRP 2*, derived from our two-index CVRP formulation, which optimizes the worst-case scenario. Then we are concerned with determining the worst-case scenario. In *Strategy 1*, the worst-case scenario takes the maximum demand values, which sometimes leads to an infeasible problem. Therefore we introduced *Strategy 2* based on an idea to satisfy as many demands as possible with respect to the fleet size and vehicle capacities. Hence, in contrast to *Strategy 1*, it always produces an appropriate solution, which reduces unmet demand, even though it is not always strictly robust. In order to solve large instances, we present our effective iterative solution method for solving RVRP by MIP solver, which iteratively improves an initial feasible solution by replacing of some arcs with other ones, obtained by exact solution of the simpler problem. Our experiments illustrated that the proposed strategy could be an useful method for solving RVRP under demand uncertainty when the worst instance is infeasible.

A possible future research is to extend this study to perform a robust analysis based on the Taguchi Method [22].

# References

1. Ben-Tal, A., Nemirovski, A.: Robust convex optimization. Math. Oper. Res. **23**(4), 769–805 (1998)
2. Bertsimas, D., Sim, M.: Robust discrete optimization and network flows. Math. Program. Ser. B **98**(1–3), 49–71 (2003)
3. Borčinová, Z., Peško, Š.: New exact iterative method for the capacitated vehicle routing problem. Commun. Sci. Lett. University of Žilina **18**(3), 19–21 (2016)
4. Borčinová, Z.: Two models of the capacitated vehicle routing problem. Croatian Oper. Res. Rev. **8**(2), 463–469 (2017)
5. Borčinová, Z., Peško, Š.: The maximum feasible scenario approach for the capacitated vehicle routing problem with uncertain demands. In: Proceedings of the 9th International Conference on Operations Research and Enterprise Systems, ICORES 2020, pp. 159–164 (2020)
6. Clarke, G., Wright, J.W.: Scheduling of vehicles from a central depot to a number of delivery points. Oper. Res. **12**(4), 568–581 (1964)
7. Dantzig, G.B., Ramser, J.H.: The truck dispatching problem. Manag. Sci. **6**(1), 80–91 (1959)
8. Gendreau, M., Potvin, J.Y.: Handbook of Metaheuristics, 2nd edn. Springer, New York (2010)
9. Gounaris, C.E., Wiesemann, W., Floudas, C.A.: The robust capacitated vehicle routing problem under demand uncertainty. Oper. Res. **61**(3), 677–693 (2013)
10. Gounaris, C.E., Repoussis, P.P., Tarantilis, C.D., Wiesemann, W., Floudas, C.A.: An adaptive memory programming framework for the robust capacitated vehicle routing problem. Transp. Sci. **50**(4), 1239–1260 (2016)
11. Gurobi Optimizer Reference Manual, version 8.1, Gurobi Optimization, Inc. https://www.gurobi.com/documentation/8.1/refman/index.html. Accessed 3 June 2020
12. Han, J., Lee, C., Park, S.: A robust scenario approach for the vehicle routing problem with uncertain travel times. Transp. Sci. **48**(3), 373–390 (2013)
13. Laporte, G., Ropke, S., Vidal, T.: Heuristics for the vehicle routing problem. In: Vehicle Routing: Problems, Methods and Applications, 2nd edn., pp. 18–89. SIAM, Philadelphia (2014)
14. Laporte, G.: Fifty years of vehicle routing. Transp. Sci. **43**(4), 408–416 (2009)
15. Laporte, G.: What you should know about the vehicle routing problem. Nav. Res. Logist. **54**, 811–819 (2007)
16. Lee, C., Lee, K., Park, S.: Robust vehicle routing problem with deadlines and travel time/demand uncertainty. J. Oper. Res. Soc. **63**(9), 1294–1306 (2012)
17. Lin, S., Kernighan, B.W.: An effective heuristic algorithm for the traveling-salesman problem. Oper. Res. **21**(2), 498–516 (1973)
18. Moghaddam, B.F., Ruiz, R., Sadjadi, S.J.: Vehicle routing problem with uncertain demands: an advanced particle swarm algorithm. Comput. Ind. Eng. **62**, 306–317 (2012)
19. Ordóñez, F.: Robust vehicle routing. In: INFORMS Tutorials in Operations Research, pp. 153–178 (2010)
20. Pessoa, A., Poss, M., Sadykov, R., Vanderbeck, F.: Solving the robust CVRP under demand uncertainty. In: 7th International Workshop on Freight Transportation and Logistics, ODYSSEUS, Calgliari, Italy, June 2018 (2018)
21. The Python Language Reference. Python Software Foundation. https://docs.python.org/3/. Accessed 3 June 2020
22. Roy, R.K.: A Primer on Taguchi Method. Reinhold International Company Ltd., 11 New Lane, London EC4P4EE, England (1990)
23. Solano-Charris, E.L., Prins, C., Santos, A.C.: A robust optimization approach for the vehicle routing problem with uncertain travel cost. In: International Conference on Control, Decision and Information Technologies (CoDIT), pp. 098–103 (2014)

24. Sun, L., Wang, B.: Robust optimization approach for vehicle routing problems with uncertainty. Int. J. Comput. Appl. Technol. (2015)
25. Sungur, I., Ordóñez, F., Dessouky, M.: A robust optimization approach for the capacitated vehicle routing problem with demand uncertainty. IIE Trans. **40**(5), 509–523 (2008)
26. Toklu, N.E., Montemanni, R., Gambardella, L.M.: An ant colony system for the capacitated vehicle routing problem with uncertain travel costs. In: IEEE Symposium on Swarm Intelligence (SIS), pp. 32–39 (2013)
27. Toth, P., Vigo, D.: Vehicle Routing: Problems, Methods and Applications, 2nd edn. Society for Industrial and Applied Mathematics (SIAM), Philadelphia (2014)

# Selected Approximate Approaches to Robust Emergency Service System Design

Marek Kvet[✉], Jaroslav Janáček, and Michal Kvet

Faculty of Management Science and Informatics, University of Žilina, Univerzitná 8215/1, 010 26 Žilina, Slovakia
{marek.kvet,jaroslav.janacek,michal.kvet}@fri.uniza.sk

**Abstract.** When the structure of a rescue service system is newly formed or redesigned, several factors possibly affecting the system performance should be taken into account. Changes in the service center deployment following from the results of associated decision-making process are often expected to be applied for a long period and thus, the obtained result has a strategic importance. Usually, the mathematical model of the rescue service system design problem with robustness incorporation takes a complicated structure. This fact disables effective solving of the problem by the branch-and-bound method due to its bad convergence. In this paper, we provide the reader with a short spectrum of our own approximate modelling techniques, which can be used to obtain a sufficient solution of the $p$-location problem by means of integer programming in acceptably short computational time. Theoretical explanation of suggested methods is here accompanied with the computational study performed with real-sized problem instances.

**Keywords:** Emergency medical service · $p$-location problem · Rescue system robustness · Radial formulation · Approximate approaches

## 1 Introduction

This paper is focused on special class of discrete network location problems, which are solved under uncertainty following from randomly occurring failures in the transportation network, through which the associated service is provided from suitably located service centers [3, 11, 30, 32, 33]. The emergency medical service system design problem is a challenging task for both system designers and operational researchers. As the first ones search for tools, which enable to obtain service center deployment satisfying future demands of the system users in case of emergency, the second ones face the necessity of completing the solving tools. Emergency service system efficiency may be considerably influenced by more or less appropriate locations of the service centers, which send emergency vehicles to satisfy service requests at the system users' locations. The number of service providing centers must be limited due to economic and technological restrictions. Obviously, the number of ambulance vehicles is finite and staff capacities are also restricted [1, 2, 4, 10, 16, 31].

© Springer Nature Switzerland AG 2022
G. H. Parlier et al. (Eds.): ICORES 2020/2021, CCIS 1623, pp. 56–82, 2022.
https://doi.org/10.1007/978-3-031-10725-2_4

In this paper, we deal with such system design objective, which represents the service accessibility of an average user. Then, the emergency service system design can be tackled as the weighted p-median problem. Any service center is assumed to have enough capacity to serve all assigned users and thus, each system user can be serviced from the nearest located service center. Host of similar models assume that serviced population is concentrated to a finite number of dwelling places of the considered area. Frequency of the demand occurrence at a given place is proportional to the number of inhabitants of the given town or village.

The necessity to solve large problem instances has led to the development of several effective solving techniques. The first group of approaches contains the exact methods based on the branch and bound principle. Many of them use a specific model reformulation, in which so-called radial approach is applied [1, 9, 12, 19]. One of the biggest disadvantages of the exact methods consists in their almost unpredictable demands, which usually make the solving process extremely time consuming. Simultaneously, many different approximate methods have been developed. Currently, the main attention is paid to various metaheuristic approaches, i.e. genetic algorithms, scatter search, path-relinking method and many others [7, 8], the aim of which can be specified as a task of obtaining a good solution in acceptably short computational time.

When the emergency service system is designed, the designer must take into account that the traversing time between a service center and the affected user might be impacted by various random events caused by weather or traffic, and possible failure of a part of critical infrastructure should be taken into account. In other words, the system resistance to such critical events should be included into the decision-making process. Most of available approaches to increasing the system resistance [13, 14, 22, 23] are based on making the design resistant to possible failure scenarios, which can appear in the associated transportation network as a consequence of random failures due to congestion, disruptions or blockages. Thus, a finite set of the failure scenarios is considered and each individual scenario is characterized by particular time distances between the users' locations and possible service center locations.

The most commonly used objective function in the above mentioned weighted p-median problem consists in minimizing the time service accessibility for an average user [16, 28], i.e. min-sum objective function is minimized subject to associated solution feasibility constraints, when the design should be obtained by using common mathematical programming tools.

The system resistance to randomly occurring detrimental events described by the set of scenarios is usually ensured so that the service center deployment has to comply with all specified scenarios. The usual way of taking into account all scenarios is based on minimizing the maximal objective function of the individual instances corresponding with particular scenarios [30, 32]. It means that instead of the min-sum objective function used in the standard weighted p-median problem, the min-max objective function is used here as a design quality level. The min-max link-up constraints and the cardinality of the scenario set significantly change the model structure. They represent an undesirable burden in any integer-programming problem due to bad convergence of the branch-and-bound method inside most available IP-solvers. Thus, alternative approximate approaches to the scenario based robust design constitute a big challenge to family of operational researchers and professionals in applied informatics.

In this paper, we present several attempts to mastering the complexity of the above scenario based robust design problem and provide the reader with a short spectrum of our own approximate modelling techniques, which can be used to obtain a sufficient solution of the $p$-location problem by means of integer programming in acceptably short computational time. All suggested and presented approaches are based on so-called radial formulation of the problem, which proved its usefulness in previous research aimed at effective solving of large instances of the weighted $p$-median problem without the concept of system robustness [9, 12, 19]. Theoretical explanation of suggested methods is here accompanied with the computational study performed with real-sized problem instances.

This book chapter represents an extended version of the paper presented at ICORES 2020 conference. The novelty of this extended paper consists in the content and structure of the paper. The conference paper was aimed directly at only one approximate solving technique for robust service system design problem, which makes use of so-called system reengineering. The original idea of system reengineering comes from previous research reported in [25]. It must be noted that any kind of system robustness was studied in mentioned literature. In the extended paper, we provide the readers with a short spectrum of approaches, which use different assumptions. A fast algorithm suggested in [20] inspired the method proposed in the conference paper, but the inner part of the algorithm was replaced by mentioned reengineering approach. In addition, the sense of system robustness is here discussed. We have added a separate section in which we analyze the possible effects of various adverse events on system performance and service delivery efficiency. Then, a short example is shown to explain the difference between basic and robust service center deployment. Another difference between the papers consists in the fact, that the extended version of the paper consists a preliminary computational study, in which the complexity and time demands of the exact mathematical model are studied. The conference paper di not contain such an analysis. Reported results confirm the necessity of approximate solving approaches development. Some of case studies and tables are taken from our previous research publications, but the reason is to provide the readers with a complex view on system robustness and to bring the possibility of comparing different developed approaches. The mathematical models were developed by the authors for this study, by the inspiration was taken and some partial approaches were taken from previous research. As an example we can mention the reengineering approach, which was developed not for solving the robust service system design problem, but in this study, this concept was incorporated into an effective and fast algorithm.

The remainder of this paper is organized as follows. The next section is devoted to the explanation of system robustness. A short example shows how the unexpected events may influence the resulting service center deployment and why it is important to consider such scenarios in the modelling phase of service system designing. Section 3 contains the exact mathematical model of robust emergency service system design problem. The necessity of approximate solving techniques is here discussed and the computational time demands of the exact model are reported. The fourth section is focused on robustness evaluation possibilities. In the fifth section, we summarize various approximate approaches and provide the reader with a short list of references. Next two sections are devoted to our developed approximate solving approaches. One of them consists in making use of the useful features of the radial formulation in solving the min-sum $p$-location problem. The second suggested method is based on so-called system reengineering, which considers also current service center deployment and brings only such changes that would be acceptable for service providers. After theoretical explanation of all studied methods, there is an overview of performed numerical experiments and the section yields also brief comparative analysis of designed service center deployments based on different values of reengineering parameters. Finally, the conclusion summarizes obtained findings and contains possible directions of further research.

## 2 Sense of System Robustness

The main goal of this section is to explain the reasons that led to the idea of rescue system robustness. Generally, when an emergency service system or any other form of public service system is newly formed or redesigned, some standard conditions on the network, through which the service is provided, are usually expected. It means that no unexpected problems or delays in service delivering are assumed. However, this assumption is not correct and it does not hold to all types of service systems. It must be realized that the emergency service system runs continuously 24 h a day, 365 days a year. Logically, many different undesirable effects that may occur randomly anywhere and anytime can negatively affect the system performance. Such critical adverse events represent uncertainty, the source of which can be divided into two main groups:

1. **Endogeneous (Internal) Causes:** These detrimental events come from the service system itself. At the time of requesting the rescue vehicle to depart, the staff may be temporarily unavailable due to the provision of a service to another patient. To incorporate such situations into the decision-making process, the concept of so-called generalized disutility has been introduced [18, 24, 26] and in many others. This model extension enables to cover more demands of system users, which can arise almost simultaneously, and furthermore, the nearest located service center does not need to have sufficient capacity to solve all the assigned demands at the same time. Of course, mentioned model generalization brings some difficulties into the solving process of the problem. Since the concept of generalized disutility does not represent the main research topic of this paper, we will restrict ourselves only on the second group of critical events that may influence the efficiency of service providing.

2. **Exogeneous (External) Causes:** This group of problems includes various adverse events on transportation networks under the influence of daytime (morning and afternoon rush hours), weather, or technical condition of roads. Such events are independent on the designed system, but they can have negative impact on its performance. Such critical events are usually described by separate distance or time matrices and thus, they form so-called detrimental scenarios [30, 32, 34, 35]. Then, the goal of the solving process is to find such service center deployment, which minimizes mentioned negative impact of the detrimental scenarios.

Another aspect that should be taken into account consists is the time perspective. Changes in the service center deployment following from the results of the associated decision-making process are often expected to be applied for a long period and thus, the obtained result has a strategic importance. Therefore, the system robustness should be considered. System robustness can be understood as a task of making the designed system resistant to random detrimental events at the maximal possible level. Let us show a simple practical demonstration.

Simple examples of various factors that may negatively affect the service system operating efficiency and cause heavy traffic on the associated transportation network are shown in Fig. 1.

**Fig. 1.** Uncertainty following from randomly occurring failures in the network [21].

The following practical example of system robustness was originally published in [21]. As it was already mentioned, randomly occurring detrimental events should be considered in the decision making process, i.e., in the strategic phase of the system designing. The following figure shows, how the elongation of one network arc affects the optimal deployment of two service centers. The small numbers placed by the network nodes represent the number of demands for service in particular network nodes. The left part of Fig. 2 represents the standard conditions and the associated optimal deployment of two service centers obtained by solving the weighted 2-median problem. The right part of Fig. 2 contains the same network graph with one elongated arc caused by a detrimental event. We can observe that the depicted arc evaluation significantly changes the optimal solution of the problem and the resulting system designs differ.

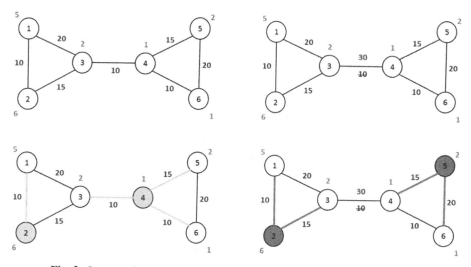

**Fig. 2.** Impact of a detrimental event on the optimal center deployment [21].

In practical applications of the robust service system design, there are usually more disjoint detrimental events and much bigger networks. Thus, finding the optimal service center deployment concerning system robustness represents a challenging task for the experts in the operations research field.

To incorporate system resistance into the associated mathematical model, a finite detrimental scenario set must be formed to cover the most commonly expected disruptions in service deliveries. This way, the associated public service system can be designed by minimization of the highest detrimental impact of the individual scenarios.

## 3  Exact Mathematical Model

The standard approach to emergency service system design usually leads to formulation of a min-sum location problem [1, 6, 16, 31], in which the average system accessibility for users (average response time) is minimized. Unfortunately, the robust system design gets more complicated structure and the problem is formulated as a min-max model bringing some difficulties into the computational process [21].

The robust emergency service system design problem using the radial formulation can be described by the following denotations. Let symbol $J$ denote the set of users' locations and let symbol $I$ denote the set of possible service center locations. We denote by $b_j$ the number of users sharing the location $j$. In some cases, the number $b_j$ could express directly the estimated number of demands for service, but it is generally assumed that the number of ambulance vehicle arrivals to the location $j$ is proportional to the population density in this location. To solve the problem, $p$ locations must be chosen from $I$ so that the maximal scenario objective function value is to be minimal. The objective function value of an individual scenario is defined as a sum of users' distances from the location of the service center providing them with service multiplied by $b_j$.

Furthermore, let symbol $U$ denote the set of possible failure scenarios. This set contains also one specific scenario called *basic scenario*, which represents standard conditions in the associated transportation network. The integer distance between locations $i$ and $j$ under a specific scenario $u \in U$ is denoted by $d_{iju}$. Even if the radial model is originally suggested for integer distance or time values only, the used principle enables us to adjust the model also for real values without any big problems.

The decisions, which determine the structure of the rescue service system, are modeled by decision variables $y_i$ for $i \in I$. The variable $y_i \in \{0, 1\}$ models the decision on service center location at the place $i \in I$ by the value of 1 if a service center is located at $i$ and by the value of 0 otherwise. In the robust problem formulation, the variable $h$ as the upper bound of the objective function issues following the individual scenarios is used. To formulate the radial model, the integer range $[0, v]$ of all possible distances of the matrices $\{d_{iju}\}$ is partitioned into zones according to [12, 19]. The value of $v$ is computed according to the expression (1).

$$v = \max\{d_{iju} : i \in I, j \in J, u \in U\} - 1 \tag{1}$$

The zone $s$ corresponds to the interval $(s, s + 1]$. Furthermore, auxiliary zero-one variables $x_{jus}$ for $j \in J$, $u \in U$ and $s = 0 \ldots v$ need to be introduced. The variable $x_{jus}$ takes the value of 1, if the distance of the user at $j \in J$ under the scenario $u \in U$ from the nearest located center is greater than $s$ and it takes the value of 0 otherwise. Then the expression $x_{ju0} + x_{ju1} + x_{ju2} + \ldots + x_{juv}$ constitutes the distance $d_{ju*}$ from user location $j$ to the nearest located service center under scenario $u \in U$. Similarly to the set covering problem, let us introduce a zero-one constant $a_{iju}^s$ under scenario $u \in U$ for each $i \in I$, $j \in J$, $s \in [0 \ldots v]$. The constant $a_{iju}^s$ is equal to 1, if the disutility $d_{iju}$ between the user location $j$ and the possible center location $i$ is less than or equal to $s$, otherwise $a_{iju}^s$ is equal to 0. Then the model of the robust system design problem can be formulated as follows.

$$\text{Minimize} \quad h \tag{2}$$

$$\text{Subject to:} \quad x_{jus} + \sum_{i \in I} a_{iju}^s y_i \geq 1 \quad \text{for } j \in J, \ u \in U, \ s = 0, 1, \ldots, v \tag{3}$$

$$\sum_{i \in I} y_i = p \tag{4}$$

$$\sum_{j \in J} b_j \sum_{s=0}^{v} x_{jus} \leq h \quad \text{for } u \in U \tag{5}$$

$$y_i \in \{0, 1\} \quad \text{for } i \in I \tag{6}$$

$$x_{jus} \in \{0, 1\} \quad \text{for } j \in J, \ u \in U, \ s = 0, 1, \ldots, v \tag{7}$$

$$h \geq 0 \tag{8}$$

The objective function (2) represented by single variable $h$ gives an upper bound of all objective function values corresponding to the individual scenarios. The constraints (3) ensure that the variables $x_{jus}$ are allowed to take the value of 0, if there is at least one center located in radius $s$ from the user location $j$ and constraint (4) limits the number of located service centers by $p$. The link-up constraints (5) ensure that each perceived disutility (time or distance) is less than or equal to the upper bound $h$. The obligatory constraints (6), (7) and (8) are included to ensure the domain of the decision variables $y_i$, $x_{jus}$ and $h$.

While the weighted $p$-median problem used for simple emergency medical system design without the robustness consideration is easily solvable, the robust design formulated by the model (2)–(8) often causes computational difficulties due to the model size and its complicated structure. Despite the fact that the radial formulation may accelerate the associated solving process [12, 19], such model simplification may not help. Obviously, each additional detrimental scenario rapidly increases the number of decision variables and structural constraints as well. Secondly, the optimization criterion (2) in combination with the link-up constraints (5) considerably slow down the convergence of the branch-and-bound process to the optimal solution. Mentioned weakness was verified in a small computational study, which is reported in the following subsection.

## 3.1 Complexity of the Exact Model and Its Computational Time Demands

It is natural, that the min-max model (2)–(8) has more complicated structure than the simple weighted $p$-median problem with a min-sum optimization criterion. That is why we expect that the min-max model size together with its bad structure can cause computational difficulties of the branch-and-bound method, especially with increasing number of detrimental scenarios. To verify this hypothesis, a small portion of preliminary experiments was performed.

To get the optimal solution of the problem (2)–(8), the optimization software FICO Xpress 8.3 (64-bit, release 2016) was used. The parameters of the computer were Intel Core i7 5500U 2.4 GHz processor and 16 GB RAM. As far as the benchmark set is concerned, we took the road network of eight Slovak self-governing regions, in which real emergency health care system is implemented. The set of used problem instances contains the regions of Bratislava (BA), Banská Bystrica (BB), Košice (KE), Nitra (NR), Trenčín (TN), Trnava (TT) and Žilina (ZA). The sets $I$ and $J$ represent all cities and villages of each region. The coefficients $b_j$ were equal to the number of inhabitants rounded up to hundreds. The parameters of individual benchmarks are summarized in the left part of Table 1. The right part of the table contains the exact solutions of the robust rescue system design problem described by the model (2)–(8). The objective function value (2) is reported in the column denoted by $h$ and the computational time in seconds is given in the column denoted by $CT$ [s].

It must be noted that one of the most important aspects of robust service system designing consists in creating an effective set of scenarios making possible to test system robustness regarding the distribution of detrimental events on relevant transportation network graph arcs. Here, many different ways can be used to create the set of scenarios. Despite the easiest random process, a specific analysis of a network graph [5, 15, 17, 27] can be applied to obtain a suitable scenario set.

Due to the lack of common benchmarks for study of robustness, the scenarios used in our computational study were created in the following way. We selected 25 percent of matrix rows so that these rows correspond to the biggest cities concerning the number of system users. Then we chose randomly from 5 to 15 rows from the selected ones and the associated disutility values in the chosen rows were multiplied by the randomly chosen constant from the range 2, 3 and 4. The rows, which were not chosen by this random process, stay unchanged. This way, ten different scenarios were generated for each self-governing region. The scenarios represent the consequence of fatal detrimental events, when some time-distances are several times elongated.

**Table 1.** Size of tested benchmarks and the exact solutions of the robust system design for self-governing regions of Slovakia with ten detrimental scenarios [20].

| Region | $|I|$ | $p$ | $CT$ [s] | $h$ |
|--------|-------|-----|----------|-----|
| BA | 87 | 9 | 28.7 | 25417 |
| BB | 515 | 52 | 1063.1 | 18549 |
| KE | 460 | 46 | 1284.2 | 21286 |
| NR | 350 | 35 | 2017.6 | 24193 |
| PO | 664 | 67 | 1180.4 | 21298 |
| TN | 276 | 28 | 264.1 | 17524 |
| TT | 249 | 25 | 433.8 | 20558 |
| ZA | 315 | 32 | 1229.7 | 23004 |

Mentioned demands of the exact model (2)–(8) can be observed also in Fig. 3, which was published in [20]. The figure shows the dependence of average computational time on the cardinality of the scenario set. Obviously, the exact approach does not hold for large problem instances because of their extremely high computational time demands.

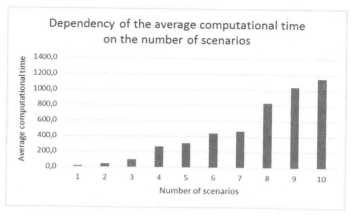

**Fig. 3.** Dependence of the average computational time on the number of scenarios [20].

As we can see in Table 1 and in Fig. 3, the exact approach to the robust rescue service system design with ten detrimental scenarios is quite time-demanding. Much worse situation occurs, when the cardinality of the scenario set doubles. The following Table 2 contains the results of the same problem instances as before with the only difference consisting in usage of 20 detrimental scenarios for each benchmark instead of 10. Additional scenarios were generated in the same way as before. Another difference is in the computational time, which was limited to 8 h. In the problems, which could not be solved to optimality by given time threshold, the best-found solution $h$ together with the available lower bound $LB$ is reported.

**Table 2.** Results of numerical experiments for 20 scenarios and 8 h of computational time.

| Region | $|I|$ | $p$ | $CT$ [s] | $h$ | $LB$ |
|--------|-----|-----|----------|-----|------|
| BA | 87 | 9 | 174.1 | 27720 | |
| BB | 515 | 52 | 28810.9 | 18830 | 18754 |
| KE | 460 | 46 | 19239.2 | 21959 | |
| NR | 350 | 35 | 28803.3 | 24584 | 24537 |
| PO | 664 | 67 | 28810.1 | 21603 | 21556 |
| TN | 276 | 28 | 15369.8 | 17775 | |
| TT | 249 | 25 | 1144.8 | 21091 | |
| ZA | 315 | 32 | 4853.9 | 23230 | |

Based on reported results, we can conclude that the exact approach to the robust rescue service system design is very time-consuming and thus, approximate solving and modelling techniques need to be developed.

## 4  Robustness Evaluation

The main goal of robust service system design is to make the system resistant to randomly occurring failures on the associated transportation network. To evaluate the gauges of robustness, several additional denotations are needed. As before, let $U$ denote the set of all considered failure scenarios, which contains also the *basic scenario*. The *basic scenario* corresponds to the standard conditions on the associated network. Let $\mathbf{y}$ denote the vector of location variables $y_i$; $i \in I$. Let $\mathbf{y}^b$ correspond to the basic system design, i.e. the solution of a simple weighted $p$-median problem, in which only the *basic scenario* is taken into account. Let $f^b(\mathbf{y})$ denote the associated objective function value. Similarly, let $\mathbf{y}^r$ denote the solution of the model (2)–(8), which brings the robust system design. Finally, the objective function (2) will be denoted by $f^r(\mathbf{y})$. The *price of robustness* (*POR*) expresses the relative increment (additional cost) of the basic scenario objective function, when $\mathbf{y}^r$ is applied instead of the optimal solution $\mathbf{y}^b$ obtained for the basic scenario. Its value is defined by (9).

$$POR = \frac{f^b(\mathbf{y}^r) - f^b(\mathbf{y}^b)}{f^b(\mathbf{y}^r)} \tag{9}$$

The *price of robustness* expresses the "bill" for making the system robust, but it does not express what we gain by such application of such solution. Therefore, we introduce also a coefficient called *gain of robustness* (*GOR*) expressed by (10).

$$GOR = \frac{f^r(\mathbf{y}^b) - f^r(\mathbf{y}^r)}{f^r(\mathbf{y}^r)} \tag{10}$$

This coefficient evaluates the profit following from applying the robust solution instead of the standard one in the worst case ignoring the detrimental scenarios [21].

## 5  Short Overview of Recent Approximate Approaches

The preliminary computational study with the exact model (2)–(8) reported in Sect. 3.1 confirmed the necessity of various alternative heuristic approaches to be developed. In this section, we provide the reader with a brief overview of possible ways to comply with the model complexity and bad convergence of the branch-and-bound method applied on the model (2)–(8) with the link-up constraints (5) linking the individual scenario objective functions up to their common upper bound $h$. We present here several approaches based on the radial formulation, which have been recently developed.

The first suggested approach follows the Lagrangean relaxation of the link-up constraints (5). Each of these constraints is associated with a non-negative Lagrangean multiplier $\lambda_u$ and the sub-gradient method is used to set up suitable values of the multipliers. The original problem (2)–(8) turns into minimization of (11) subject to (2), (4), (6), (7) and (8).

$$\text{Minimize} \quad h + \sum_{u \in U} \lambda_u \left( \sum_{j \in J} \sum_{s=0}^{v} b_j x_{jsu} - h \right) = h \left( 1 - \sum_{u \in U} \lambda_u \right) + \sum_{u \in U} \lambda_u \sum_{j \in J} \sum_{s=0}^{v} b_j x_{jsu} \quad (11)$$

Obviously, the problem (11), (3), (4), (6), (7) and (8) has a feasible solution only for such setting of Lagrangean multipliers, where their sum is less than or equal to one. Having the optimal solution of the relaxed problem with arbitrary multipliers meeting the stated rule, the value of the optimal solution yields a lower bound of the optimal solution of the original problem (2)–(8). If the sum of the Lagrangean multipliers is less than one, then the optimal value of $h$ would equal to zero and it does not represents the upper bound at all. That is why we restrict ourselves on such setting of multipliers, where the sum equals to one exactly. We try to reach or at least to approximate the optimal solution of the original problem (2)–(8) by an iterative algorithm, where the Lagrangean multipliers $\lambda_u$ are recomputed in each iteration. This algorithm was introduced in [13], where also other details are discussed.

The rest of presented approaches are based on the idea of replacing the set $U$ of individual scenarios by one supplementary scenario. After this assumption, the original exact min-max model (2)–(8) turns into a simplified to the form (12)–(16).

$$\text{Minimize} \quad \sum_{j \in J} b_j \sum_{s=0}^{v} x_{js} \quad (12)$$

$$\text{Subject to:} \quad x_{js} + \sum_{i \in I} a_{ij}^s y_i \geq 1 \quad \text{for } j \in J, \ s = 0, 1, \ldots, v \quad (13)$$

$$\sum_{i \in I} y_i = p \quad (14)$$

$$y_i \in \{0, 1\} \quad \text{for } i \in I \quad (15)$$

$$x_{js} \geq 0 \quad \text{for } j \in J, \ s = 0, 1, \ldots, v \quad (16)$$

All decision variables and structural constraints take the same meaning as before. The objective function (12) minimizes the disutility perceived by an average user, what can be achieved by minimizing the sum of time or distance values perceived by all system users. The only difference consists in the fact, that the indexes $u$ over the set $U$ of scenarios used in the variables and constraints of the previous exact mathematical model (2)–(8) are not necessary.

As it was discussed in previous sections, simple weighted p-median problem described by the model (12)–(16) is much better solvable than the original min-max problem formulated by expressions (2)–(8).

Generally, the constant $a_{ij}^s$ for $i \in I, j \in J, s \in [0 \ldots v]$ is equal to 1, if the disutility $d_{ij}$ perceived by any user located at the location $j$ from the possible center location $i$ is less than or equal to $s$, otherwise $a_{ij}^s = 0$. The disutility $d_{ij}$ may be defined in many different ways. It may represent time necessary for service delivering or the distance, which is needed to overcome in order to get the demand satisfied.

The fuzzy approach [14] does not deal with the set of given crisp scenarios as proposed in the model (2)–(8), but it describes the individual uncertain values, e.g. the value of perceived disutility, by a range of possible values together with a measure of relevance of the individual values from the range. The measure is called *membership function* and the function is defined on the whole set of real numbers and maps this definition range on the interval [0, 1]. The value of one is assigned to the real values that belong to the fuzzy value at the highest level and the value of zero is intended for real values outside the range of the fuzzy value. A particular fuzzy set defined on real numbers membership function of which it satisfies given conditions is called *fuzzy number*. We restrict ourselves to a special type of the fuzzy sets intended for expressing that a processed uncertain disutility value is greater than or equal to a given real value $d_{ij-min}$ with tolerance $d_{ij-Max} - d_{ij-min}$. Here, $d_{ij-min} = min\{d_{iju}: u \in U\}$ and $d_{ij-Max} = max\{d_{iju}: u \in U\}$ (Fig. 4).

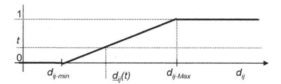

**Fig. 4.** The membership function of a fuzzy set $\underline{d}_{ij}$.

In this fuzzy approach, the constant $a_{ij}^s$ must be defined for each pair $[i, j]$ and given level of satisfaction $t \in [0, 1]$. The constant $a_{ij}^s(t)$ is equal to 1, if the disutility $\underline{d}_{ij}(t)$ between the user location $j$ and the possible center location $i$ is less than or equal to $s$, otherwise $a_{ij}^s(t) = 0$. After these preliminaries, we can solve the problem described by the model (12)–(16) with fuzzy coefficients by iterative Tanako-Asai's approach [29, 36], in which we use the bisection method to search for the highest level of satisfaction $t^*$, for which the problem has a feasible solution. Obviously, one of the problems connected with the fuzzy approach consists in a proper way of fuzzy number computation, which may logically affect the resulting level of satisfaction and the service center deployment as well. The disadvantage of previously introduced approach consists in the fact that it does not process the complete set $U$ in a sufficient way. The minimum and maximum of each disutility over the set $U$ does not bring any information about other scenarios, which should be also considered. Therefore, we have suggested also another approach.

This newly suggested approach is based on a convex combination of individual scenarios. Similarly to the Lagranean relaxation described above, the set of nonnegative multipliers $\lambda_u \geq 0$ for $u \in U$ is used to define the resulting value of each disutility $\underline{d}_{ij}$ perceived by any user located at $j$ from the possible center location $i$. It must be noted that the coefficients $\lambda_u$ have to satisfy the condition that their sum equals to one. These multipliers can also represent the probability distribution of the individual scenarios. Then, the disutility values $\underline{d}_{ij}$ can be expressed according to (17) as follows.

$$\underline{d}_{ij}(\lambda) = \sum_{u \in U} \lambda_u d_{iju} \quad for \ i \in I, \ j \in J \tag{17}$$

The values of $\underline{d}_{ij}$ depend on the settings of multipliers $\lambda_u$. After these preliminaries, the covering constants $a_{ij}^s$ are computed and the radial model (12)–(16) is iteratively solved. In each iteration, the multipliers $\lambda_u$ are redefined based on the obtained solution according to certain rules [22, 23]. The radial model is being solved, while better solution of the system design keeps being obtained or until other stopping criteria are met.

The main advantage of suggested heuristic approaches consists in the fact that the approximate methods perform in order faster than the original standard exact approach, which is formulated by the model (2)–(8). The difference is caused by the fact, that the approximate approaches do not process all the detrimental scenarios and avoid introducing the min-max link-up constraints, which generally cause bad convergence of the branch-and-bound method.

As far as the solution accuracy is concerned, the quality of obtained results is usually strongly dependent on the initial settings of parameters or on the way of fuzzy numbers definition. If we had to choose the best heuristic method out of mentioned approaches, we would pick the convex approach, because it takes into account the complete set $U$ of scenarios and does not reflect only to some selected values of disutility. Therefore, we will concentrate our effort on development of such algorithm that would find a good solution in a short time without introducing specific parameters.

## 6 Fast Algorithm with Fixed Location Variables

The main reason for developing approximate solving approaches to robust rescue system design problem is to overcome bad solvability of the original problem described by the model (2)–(8) caused by its structure and big size. It must be realized that each additional detrimental scenario magnifies the model size proportionally to the cardinality of the set $I$.

Presented algorithm is based on the assumption that not only the cardinality of the set $U$, link-up constraints (5) and the min-max objective function (2), but also the number of location variables $y_i$ play an important role and make the model extremely time-consuming for the associated solving method. The main idea of this effective and fast algorithm consists in the following expectation. If we were able to make the decision problem smaller by fixing some variables $y_i$ either to the value 0 or 1, then we could considerably accelerate the solving process. Thus, the original exact model can be replaced by series of smaller sub-problems to find suitable variables $y_i$ for fixing. After that, we can solve a smaller min-max robust rescue system design problem. Mentioned adjustment turns the original exact approach to an approximate solving technique. To formulate a new mathematical model with mentioned adjustment, three new disjoint sets must be defined in such a way, that their union brings the original set of all possible service center locations $I$. Let the set $F_1$ contain such indexes $i \in I$, for which $y_i$ must equal to 1, i.e. a service center must be located in each element of $F_1$. Analogically, the set $F_0$ is formed by those center locations, in which a center cannot be located. Here, $y_i = 0$ for each $i \in F_0$. Finally, let the indexes of variables $y_i$, which do not have assigned their values yet, form a finite set denoted by $V$. Mathematically, the set $V$ can be defined as $V = I - F_1 - F_0$. Furthermore, it is enough to define variables $y_i$ only for the locations from the set $V$, because the decisions of other service center locations are already made by the sets $F_0$ and $F_1$. Based on these preliminaries, the obligatory constraints (6) can be replaced by (18) and the constraints (4) can take the form of (19). In the expression (19), the symbol $p_{rem}$ is used to denote the number of service centers, which remain to the original number of all located service centers $p$. It means, $p_{rem} = p - |F_1|$. Obviously, it is assumed that the set $F_1$ contains less than $p$ elements.

$$y_i \in \{0, 1\} \quad for \ i \in V \tag{18}$$

$$\sum_{i \in V} y_i = p_{rem} \tag{19}$$

Similarly, it is not necessary to define the coefficient $a^s_{iju}$ for each $i \in I$ as above. It is enough to define it for each element $i$ from $V \cup F_1$. Then, the constraints (3) can be redefined to (20).

$$x_{jus} + \sum_{i \in V} a^s_{iju} y_i + \sum_{i \in F_1} a^s_{iju} \geq 1 \quad for \ j \in J, \ u \in U, \ s = 0, 1, \ldots, v \tag{20}$$

After mentioned model adjustment suggestions, the approximate radial model for robust rescue system design takes the form of (21)–(27). The meaning of decision variables stays unchanged.

$$Minimize \quad h \tag{21}$$

$$Subject \ to: \quad x_{jus} + \sum_{i \in V} a^s_{iju} y_i + \sum_{i \in F_1} a^s_{iju} \geq 1 \quad for \ j \in J, \ u \in U, \ s = 0, 1, \ldots, v$$

$$\tag{22}$$

$$\sum_{i \in V} y_i = p_{rem} \tag{23}$$

$$\sum_{j \in J} b_j \sum_{s=0}^{v} x_{jus} \leq h \quad for \ u \in U \tag{24}$$

$$y_i \in \{0, 1\} \quad for \ i \in V \tag{25}$$

$$x_{jus} \in \{0, 1\} \quad for \ j \in J, \ u \in U, \ s = 0, 1, \ldots, v \tag{26}$$

$$h \geq 0 \tag{27}$$

Filling up the sets $F_1$ and $F_0$ can be done in the following way. Let us introduce a set $I^u_1$ for each detrimental scenario $u \in U$. The set $I^u_1$ contains the solution of a simple weighted $p$-median problem obtained for each scenario $u$ separately. The set $I^u_1$ for given $u \in U$ can be mathematically formulated by (28).

$$I^u_1 = \{i \in I : y_i = 1\} \quad for \ u \in U \tag{28}$$

The mentioned location problem solved for each scenario $u \in U$ to define the set $I_1^u$ can be formulated either by the model (2)–(8), in which the scenario set $U$ contains only the scenario $u$, or we can solve a simple weighted $p$-median problem making use of the above introduced constants and variables. The radial formulation for a given scenario $u$ takes the form of (12)–(16). In the model (12)–(16), we work only with one scenario $u$ from the scenario set $U$ and therefore, we do not use the index $u$ in variables $x_{js}$.

As shown in [12, 19], this model can be solved to optimality in a very short time thank to good characteristics of the covering problem. After computing the set $I_1^u$ for each $u \in U$, we can fill the sets $F_1$ and $F_0$ according to (29) and (30) respectively.

$$F_1 = \bigcap_{u \in U} I_1^u \tag{29}$$

$$F_0 = I - \bigcup_{u \in U} I_1^u \tag{30}$$

Finally, the complete approximate algorithm for rescue system designing can be summarized according to [20] by the following steps:

1. Determine the set $I_1^u$ for each detrimental scenario $u \in U$ by solving the model (12)–(16).
2. Fill up the sets $F_1$ and $F_0$ according to (29) and (30).
3. Solve the model (21)–(27) to obtain the resulting system design.

Suggested approximate algorithm can be easily implemented in any common optimization environment without the necessity of special software tool development. It is based on two integer programming problems, which can be solved by the branch-and-bound or similar method. Another big advantage of presented approach consists in its independence on any parameters, which are often sensitive to their values and may directly affect not only the performance characteristics (computational time), but also the obtained results. Thus, this approximate solving method can be considered very useful.

## 7 Reengineering Approach

The problem of many optimization approaches applied on real systems consists in the fact that the resulting service center deployment may bring too big changes in service center deployment, which would be hardly acceptable for service providers. To avoid such potential complications, the concept of system reengineering has been introduced [24–26]. Its most important feature lies in considering current service center deployment and performing only acceptable relocations satisfying given restrictions. In the approximate robust system design algorithm suggested in Sect. 6, the reengineering concept can be incorporated into the solving process in the way of filling the sets $F_0$ and $F_1$ respectively [25]. Let us now remind the basis of system reengineering.

The basic idea follows from the analysis of current service center deployment, which may not be optimal due to changing demands and development of the underlying transportation network. To explain the problem in more details, consider the simple example depicted in Fig. 5. We assume that the left graph represents current deployment of four service centers marked by blue color. All the vertices represent possible demand points. To evaluate the current center deployment, the sum of distances from each network node to the nearest located center was used as the quality criterion. Here, it takes the value of 66. If we allowed changes in current service center locations and moved a service center from the node 2 to the node 6, we would perform system reengineering and we could achieve better value of the criterion. The new system design is depicted on the right graph and its evaluation is 64. By this small example, we demonstrated the principle and goal of system reengineering.

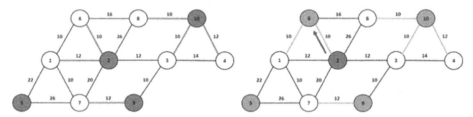

**Fig. 5.** Simple example of emergency service system reengineering [26].

To describe the problem of the system average distance or response time minimization by changing the deployment of centers, let $I$ be a finite set of all possible center locations. As above, the response time following from the distance between locations $i$ and $j$ under a specific scenario $u \in U$ is denoted as $d_{iju}$. The union of two disjoint sets of located centers $L$ and $F_1$, where $I_L$ contains $p$ centers under reconstruction and $F_1$ is the set of fixed centers, describe the current emergency service center deployment. The center locations from $I_L$ can be relocated within the set $I - F_1 - F_0$, where $F_0$ is the set of temporarily forbidden locations.

When reengineering of an emergency service system is performed, the administrator of the system may set up parameters of rules to prevent a designer of new center deployment from changes, which can be considered by system users as obnoxious. We consider two formal rules within this study. The first rule limits the total number $w$ of the centers, location of which can be changed. The second rule limits the time distance between current and newly suggested location of a service center by the given value $D$. This rule can be explained also by the following example depicted in Fig. 6. We assume that all points 1–11 represent system users and the black points 2, 3, 9 and 11 represent current service center locations.

To be able to formulate the rules in a concise way, we derive several auxiliary structures. Let $N_t$ denote the set of all possible center locations, to which the center $t \in I_L$ can be moved subject to limited length of the move. The set $N_t$ can be mathematically formulated by the expression (31).

$$N_t = \{i \in I - F_1 - F_0 : d_{ti} \leq D\} \quad \text{for } t \in I_L \tag{31}$$

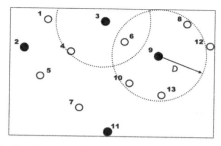

**Fig. 6.** Simple example of reengineering restrictions [26].

If we consider the example depicted in Fig. 6, we can observe that the center located at the point 9 can be moved to 6, 8, 10 and 13 or stay unchanged. Thus, the set $N_9 = \{6, 8, 9, 10, 13\}$. Similarly, $N_3 = \{3, 4, 6\}$.

Additionally, let $S_i$ denote the set of all centers, which can be moved to $i \in I - F_1 - F_0$ subject to the mentioned limitation as described by (32). Here, $S_6 = \{3, 9\}$.

$$S_i = \{t \in I_L : i \in N_t\} \quad for \ i \in I - F_1 - F_0 \tag{32}$$

Now, we introduce series of decision reallocation variables, which model the decisions on moving centers from their original positions to new ones. The variable $u_{ti} \in \{0, 1\}$ for $t \in L$ and $i \in N_t$ takes the value of one, if the service center at $t$ is to be moved to $i$ and it takes the value of zero otherwise. The following model (33)–(41) describes the problem of reengineering for given detrimental scenario $u$.

$$Minimize \quad \sum_{j \in J} b_j \sum_{s=0}^{v} x_{js} \tag{33}$$

$$Subject \ to: \quad x_{js} + \sum_{i \in I-F1-F0} a_{iju}^s y_i + \sum_{i \in F1} a_{iju}^s \geq 1 \quad for \ j \in J, \ s = 0, \ 1, \ \dots, \ v \tag{34}$$

$$\sum_{i \in I-F1-F0} y_i = p \tag{35}$$

$$\sum_{i \in L} y_i \geq p - w \tag{36}$$

$$\sum_{i \in N_t} u_{ti} = 1 \quad for \ t \in L \tag{37}$$

$$\sum_{t \in S_i} u_{ti} \leq y_i \quad for \ i \in I - F_1 - F_0 \tag{38}$$

$$u_{ti} \in \{0, \ 1\} \quad for \ t \in L, \ i \in N_t \tag{39}$$

$$y_i \in \{0, \ 1\} \quad for \ i \in I - F_1 - F_0 \tag{40}$$

$$x_{js} \geq 0 \quad for \ j \in J, \ s = 0, \ 1, \ \ldots, \ v \tag{41}$$

As the constraints (34), (35) and formulae together with used decision variables were explained in the previous section, we restrict the explanation only on the remainder of the above model. Constraint (36) limits the number of changed center locations by the constant $w$. The next constraints (37) allow moving the center from the current location $t$ to at most one other possible location in the radius $D$. Constraints (38) enable to bring at most one center to a location $i$ subject to the condition that the original location of the brought center lies in the radius $D$. These constraints also assure consistency among the decisions on move and decisions on center location. The constraints (39), (40) and (41) are used to keep the definition scope of used variables.

Having introduced and explained the service system reengineering problem, this model can be incorporated into the approximate solving algorithm for robust service system designing. If we set the parameters $w$ and $D$ of the reengineering method at chosen values from the ranges $[1 \ldots p]$ and $[1 \ldots D^{max}]$ respectively, and initialize the sets $F_1 = \emptyset$ and $F_0 = \emptyset$, then the following steps can describe the suggested solving algorithm:

1.  Determine the set $I_1^u$ for each detrimental scenario $u \in U$ by solving the reengineering model (33)–(41) with the parameters $w$ and $D$.
2.  Fill up the sets $F_1$ and $F_0$ according to (29) and (30).
3.  Solve the model (21)–(27) to obtain the resulting system design.

Let the indexes $i \in I - F_1 - F_0$, for which the location variable $y_i$ in the model (21)–(27) take the value of 1 together with the indexes $i$ from the set $F_1$ form the resulting service system deployment described by the set $I_{res}$ as defined by (42).

$$I_{res} = F_1 \cup \{i \in I - F_1 - F_0 : y_i = 1\} \tag{42}$$

The objective function value $h$ of any resulting set $I_{res}$ can be computed according to the expression (43).

$$h = f(I_{res}) = \max \left\{ \sum_{j \in J} b_j \min \{d_{iju} \ : \ i \in I_{res}\} \ : \ u \in U \right\} \tag{43}$$

## 8   Computational Study

### 8.1   Benchmarks and Preliminaries

The numerical experiments reported in this case study were performed in the optimization software FICO Xpress 8.3, 64-bit. The experiments were run on a PC equipped with the Intel® Core™ i7 5500U 2.4 GHz processor and 16 GB RAM.

Used benchmarks were derived from real emergency health care system, which was implemented in eight regions of Slovak Republic. For each self-governing region, i.e.

Bratislava (BA), Banská Bystrica (BB), Košice (KE), Nitra (NR), Prešov (PO), Trenčín (TN), Trnava (TT) and Žilina (ZA), all cities and villages with corresponding number of inhabitants $b_j$ were taken into account. The coefficients $b_j$ were rounded up to hundreds. The set of communities represents both the set $J$ of users' locations and the set $I$ of possible center locations as well. The cardinalities of these sets vary from 87 to 664 locations. The organization of the Slovak self-governing regions is depicted in Fig. 7.

**Fig. 7.** Used benchmarks – self-governing regions of Slovakia.

The parameters of individual benchmarks are summarized in the Table 1 in Sect. 3.1, in which also the exact solutions taken from previous research [21] are reported. For this computational study, 10 scenarios were generated according to the algorithm making use of random generation. The main goal of this computational study was to verify basic performance characteristics of approaches suggested in Sects. 6 and 7.

The solution accuracy is evaluated by so-called *gap*, which is defined in the following way. Let $ObjF_{ex}$ denote the objective function value of the exact solution and let us use $ObjF_{ap}$ to denote the objective function value of the approximate one. Then we can evaluate *gap* in percentage of $ObjF_{ex}$ according to the following expression (44).

$$gap = 100 * \frac{ObjF_{ap} - ObjF_{ex}}{ObjF_{ex}} \tag{44}$$

The resulting system designs are compared by *Hamming distance* of the location variables vectors. Generally, *Hamming distance HD* of two vectors $\mathbf{y}^a$ and $\mathbf{y}^b$ is defined by the expression (45). Obviously, the vectors must have the same size.

$$HD\left(\mathbf{y}^a, \mathbf{y}^b\right) = \sum_{i \in I} \left(y_i^a - y_i^b\right)^2 \tag{45}$$

## 8.2 Experiments with the Fast Algorithm with Fixed Location Variables

This chapter contains the results of performed computational study, the goal of which was to verify basic performance characteristics of suggested fast algorithm with fixed location variables. As discussed in previous sections, the exact approach to robust rescue system design formulated by the model (2)–(8) is very time-demanding and thus, approximate solving and modelling techniques need to be developed. The results obtained by performed numerical experiments are reported in the following Table 3.

Each row of the table corresponds to one solved problem instance. The objective function value computed for the resulting system design according to the expression (43) is reported in the column denoted by $h$ and the computational time in seconds is given in the column denoted by $CT$ [s]. Furthermore, the solution accuracy is evaluated by so-called *gap*, which is defined in the form of (44). Finally, the resulting system designs are compared by *Hamming distance* of the location variables vectors. These values are reported in the last column of the table denoted by $HD$.

**Table 3.** Results of the approximate approach for the self-governing regions of Slovakia [20].

| Region | $|I|$ | $p$ | Approximate solution | | | |
|---|---|---|---|---|---|---|
| | | | $CT$ [s] | $h$ | $gap$ [%] | $HD$ |
| BA | 87 | 9 | 19.2 | 25417 | 0.00 | 0 |
| BB | 515 | 52 | 84.3 | 18779 | 1.24 | 18 |
| KE | 460 | 46 | 66.4 | 21664 | 1.78 | 14 |
| NR | 350 | 35 | 65.0 | 24392 | 0.82 | 12 |
| PO | 664 | 67 | 129.6 | 21584 | 1.34 | 18 |
| TN | 276 | 28 | 62.7 | 17744 | 1.19 | 10 |
| TT | 249 | 25 | 29.5 | 21084 | 2.56 | 12 |
| ZA | 315 | 32 | 24.8 | 23359 | 1.54 | 10 |

The results reported in Table 3 clearly confirm expected useful features of suggested approximate algorithm, because it provides us with very satisfactory solutions from the viewpoint of accuracy. Furthermore, the demanded computational time is in order smaller than required by the exact approach. When analyzing the values of location variables $y_i$ for $i \in I$, we have found the assumption that obtaining better solution may be probably limited by the set $F_1$, which contains also such indexes $i \in I$ of variables $y_i$ fixed to 1, that should belong to the set $V$ instead of $F_1$. The second portion of numerical experiments verified this hypothesis. Here, the set $F_1$ keeps empty and the approximate algorithm uses only the set $F_0$. In such a case, we can expect these two consequences:

1. Obtaining better result from the viewpoint of objective function value (43), because there are less location variables fixed to their final values and thus, the solution accuracy may get better.
2. Higher computational time demands, because there are more variables, which need to get their final values.

The following portion of experiments, results of which are reported in Table 4, confirmed all mentioned expected consequences. This table has the same structure as previously reported Table 3.

**Table 4.** Results of the approximate approach applied on the self-governing regions, in which only set $F_0$ is used [20].

| Region | $|I|$ | $p$ | Approximate solution | | | |
|---|---|---|---|---|---|---|
| | | | $CT$ [s] | $h$ | $gap$ [%] | $HD$ |
| BA | 87 | 9 | 13.5 | 25417 | 0.00 | 0 |
| BB | 515 | 52 | 203.0 | 18596 | 0.25 | 12 |
| KE | 460 | 46 | 179.1 | 21341 | 0.26 | 4 |
| NR | 350 | 35 | 319.7 | 24193 | 0.00 | 0 |
| PO | 664 | 67 | 216.0 | 21311 | 0.06 | 6 |
| TN | 276 | 28 | 103.1 | 17628 | 0.53 | 8 |
| TT | 249 | 25 | 83.6 | 20768 | 1.2 | 4 |
| ZA | 315 | 32 | 142.8 | 23004 | 0.00 | 0 |

As we can see in Table 4, presented approximate algorithm, which uses only the set $F_0$, brings excellent quality of the resulting solution, which is near to the optimal one. Another useful feature consists in computational time, which keeps satisfactorily small.

## 8.3 Experiments with the Reengineering Approach

The last but not least important portion of numerical experiments was focused on the approximate solving technique described in Sect. 7, which subsequently solves the system reengineering problem for each detrimental scenario as proposed in [25].

The following series of experiments was performed with the goal to find a suitable setting of the parameters $w$ and $D$ used in the approximate approach, see the model (33)–(41) of the reengineering process. For this study, the benchmark Žilina (ZA) was used. In this portion of experiments, the parameter $p$ was set at the value 32 reported in previous tables. The maximal radius $D$ was fixed at one of the values 5, 10, 15, 20 and 25 and the maximal number $w$ of centers allowed to change their locations was set to $p/4$, $p/2$, $3p/4$, and $p$ respectively. The results of this series of experiments are summarized in Table 5.

Each row of the table corresponds to one setting of the parameters $w$ and $D$. In this portion of experiments, several characteristics were studied. The computational time in seconds is reported in the column denoted by $CT$ [s]. Since the approximate approach proved to be much faster than the exact one, the percentage save of computational time $SCT$ was computed. Here, the computational time of the exact approach was taken as the base. Furthermore, the objective function associated with the obtained service center deployment is reported in the column denoted by $Obj^{ap}$. Its value was computed according to (43). To evaluate the accuracy of suggested approximate method, the value of *gap* was also computed. It expresses the difference between the objective function values of the exact and approximate models. The objective value of the exact approach was taken as the base. The value of *gap* is reported also in percentage. Finally, the resulting service center deployments were compared in the terms of *Hamming distance* of the vectors of location variables $y_i$. This value is denoted by *HD*.

It can be seen that the lowest computational time of the approximate method was reached for the settings $w = p/4$ and $D = 5$.

**Table 5.** Detailed results of numerical experiments for the self-governing region of Žilina: computational study of the impact of individual parameters on the results accuracy [25].

| $w$ | $D$ | $CT$ [s] | $SCT$ [%] | $ObjF^{ap}$ | $gap$ [%] | $HD$ |
|---|---|---|---|---|---|---|
| 8 | 5 | 231.6 | 82.25 | 23411 | 1.77 | 14 |
| 8 | 10 | 430.9 | 66.98 | 23377 | 1.62 | 14 |
| 8 | 15 | 502.8 | 61.46 | 23236 | 1.01 | 6 |
| 8 | 20 | 516.8 | 60.39 | 23359 | 1.54 | 10 |
| 8 | 25 | 529.5 | 59.42 | 23359 | 1.54 | 10 |
| 16 | 5 | 234.8 | 82.00 | 23411 | 1.77 | 14 |
| 16 | 10 | 455.1 | 65.12 | 23377 | 1.62 | 14 |
| 16 | 15 | 516.0 | 60.45 | 23236 | 1.01 | 6 |
| 16 | 20 | 529.9 | 59.38 | 23359 | 1.54 | 10 |
| 16 | 25 | 542.6 | 58.41 | 23359 | 1.54 | 10 |
| 24 | 5 | 232.5 | 82.18 | 23411 | 1.77 | 14 |
| 24 | 10 | 434.6 | 66.69 | 23377 | 1.62 | 14 |
| 24 | 15 | 508.5 | 61.02 | 23236 | 1.01 | 6 |
| 24 | 20 | 531.3 | 59.28 | 23359 | 1.54 | 10 |
| 24 | 25 | 534.6 | 59.02 | 23359 | 1.54 | 10 |
| 32 | 5 | 231.6 | 82.25 | 23411 | 1.77 | 14 |
| 32 | 10 | 427.8 | 67.21 | 23377 | 1.62 | 14 |
| 32 | 15 | 509.2 | 60.97 | 23236 | 1.01 | 6 |
| 32 | 20 | 531.4 | 59.27 | 23359 | 1.54 | 10 |
| 32 | 25 | 534.8 | 59.01 | 23359 | 1.54 | 10 |

As the associated *gap* was acceptable, we used this setting in the third series of experiments, which was performed for each self-governing region. The obtained results are reported in Table 6, in which the same denotations as in previous tables were used. The table is divided into two sections denoted by **EXACT** and **APPROXIMATE**. The first section contains the results from Table 1 for bigger comfort of the readers. The second section contains the results obtained by the approximate approach.

**Table 6.** Comparison of the approximate approach with the system reengineering to the exact approach. Parameters of the approximate approach were set in this way: $w = p/4$, $D = 5$ [25].

|      | EXACT | | APPROXIMATE | | | |
|------|---------|----------|---------|----------|---------|------|
|      | $CT$ [s] | $ObjF^{ex}$ | $CT$ [s] | $ObjF^{ap}$ | $gap$ [%] | $HD$ |
| BA   | 52.8    | 25417    | 11.8    | 26197    | 3.07    | 4    |
| BB   | 1605.0  | 18549    | 679.8   | 18861    | 1.68    | 12   |
| KE   | 1235.5  | 21286    | 633.4   | 21935    | 3.05    | 16   |
| NR   | 11055.1 | 24193    | 274.2   | 24732    | 2.23    | 14   |
| PO   | 3078.2  | 21298    | 1601.9  | 21843    | 2.56    | 20   |
| TN   | 616.6   | 17535    | 223.1   | 17851    | 1.80    | 10   |
| TT   | 563.8   | 20558    | 152.4   | 20980    | 2.05    | 10   |
| ZA   | 1304.7  | 23004    | 231.6   | 23411    | 1.77    | 14   |

The approximate approach enables to obtain the resulting robust service center deployment in the computational time, which is much less than half of the computational time demanded by the exact approach. As far as the accuracy of the resulting solution is concerned, it can be observed that the approximate method is very satisfactory.

## 9  Conclusions

This paper was focused on advanced way of rescue service system designing. Emergency medical system plays a very important role in human life. It must be realized that any potentially bad decision made in the process of ambulance vehicle stations locating may bring not only enormous economic losses, but it can cause serious lasting consequences for human health and thus, it usually negatively affects the most vulnerable part of population. Therefore, it is necessary to consider all aspects that may have a bad impact on service delivering and efficiency of system performance.

Within this study, we paid attention to system robustness, which means its resistance to randomly occurring failures in the associated transportation network used by ambulance vehicles when delivering rescue service to system users. To incorporate system robustness into the particular integer-programming model, a finite set of failure scenarios is usually formed. Then, the optimal robust rescue system design can be obtained by minimization of the highest detrimental impact of the individual scenarios.

Since the exact mathematical model has a complicated structure, which causes bad convergence of the branch-and-bound method, several approximate approaches need to be developed. In this paper, we provided the readers with a short spectrum of our own approximate modelling techniques, which can be used to obtain a sufficient solution of the $p$-location problem by means of integer programming in acceptably short computational time. Special attention was paid to a specific model adjustment consisting in fixing some location variables and in making the decision-making problem smaller. We have suggested two different variants of mentioned approximate algorithm. One of them consists in making use of the useful features of the radial formulation in solving the min-sum $p$-location problem. The second suggested method is based on system reengineering, which considers also current service center deployment and brings only such changes that would be acceptable for service providers. Theoretical explanation of all suggested methods was accompanied with the computational study performed with real-sized problem instances.

The performed numerical experiments have confirmed the usefulness of suggested methods. They are simply implementable in any commercial IP-solver without the necessity of special knowledge in the Applied Informatics field. Based on achieved results we can conclude that we have presented very useful tools for robust system designing.

In the future, we would like to concentrate on other approximate techniques, which would enable us to reach shorter computational time under acceptable solution accuracy. Another research goal could be focused on developing such a heuristic approach, which would form the second stage of presented algorithm and could improve the obtained resulting system design. Furthermore, we could focus on goal programming approaches to robust service system design or on developing advanced solving techniques based on evolutionary or machine learning methods.

**Acknowledgment.** This work was supported by the research grants VEGA 1/0342/18 "Optimal dimensioning of service systems", VEGA 1/0089/19 "Data analysis methods and decisions support tools for service systems supporting electric vehicles" and VEGA 1/0689/19 "Optimal design and economically efficient charging infrastructure deployment for electric buses in public transportation of smart cities". This work was supported by the Slovak Research and Development Agency under the Contract no. APVV-19-0441.

# References

1. Avella, P., Sassano, A., Vasil'ev, I.: Computational study of large scale p-median problems. Math. Program. **109**(1), 89–114 (2007)
2. Brotcorne, L., Laporte, G., Semet, F.: Ambulance location and relocation models. Eur. J. Oper. Res. **147**, 451–463 (2003)
3. Correia, I., Saldanha da Gama, F.: Facility locations under uncertainty, Location Science, eds. Laporte, Nikel, Saldanha da Gama, pp. 177–203 (2015)
4. Current, J., Daskin, M., Schilling, D.: Discrete network location models. In: Drezner, Z., et al. (eds.) Facility Location Applications and Theory, pp. 81–118. Springer, Berlin (2002)
5. Czimmermann, P., Koháni, M.: Characteristics of changes of transportation performance for pairs of critical edges. Commun. Sci. Lett. Univ. Žilina **20**(3), 84–87 (2018)
6. Czimmermann, P., Koháni, M.: Computation of transportation performance in public service systems. In: IEEE Workshop on Complexity in Engineering - COMPENG 2018, pp. 1–5. Institute of Electrical and Electronics Engineers, Danvers (2018)
7. Doerner, K.F., et al.: Heuristic solution of an extended double-coverage ambulance location problem for Austria. CEJOR **13**(4), 325–340 (2005)
8. Elloumi, S., Labbé, M., Pochet, Y.: A new formulation and resolution method for the p-center problem. INFORMS J. Comput. **16**, 84–94 (2004)
9. García, S., Labbé, M., Marín, A.: Solving large p-median problems with a radius formulation. INFORMS J. Comput. **23**(4), 546–556 (2011)

10. Guerriero, F., Miglionico, G., Olivito, F.: Location and reorganization problems: the Calabrian health care system case. Eur. J. Oper. Res. **250**(3), 939–954 (2016)
11. Ingolfsson, A., Budge, S., Erkut, E.: Optimal ambulance location with random delays and travel times. Health Care Manag. Sci. **11**(3), 262–274 (2008)
12. Janáček, J.: Approximate covering models of location problems. In: Proceedings of the 1st International Conference ICAOR. Lecture Notes in Management Science, Yerevan, Armenia, pp. 53–61 (2008)
13. Janáček, J., Kvet, M.: Designing a robust emergency service system by Lagrangean relaxation. In: Mathematical Methods in Economics 2016, Liberec, pp. 349–353 (2016)
14. Janáček, J., Kvet, M.: An approach to uncertainty via scenarios and fuzzy values. Croatian Oper. Res. Rev. **8**(1), 237–248 (2017)
15. Janáček, J., Kvet, M.: Detrimental scenario construction based on network link characteristics. In: Proceedings of the 19th IEEE International Carpathian Control Conference (ICCC), Szilvásvárad, pp. 629–632 (2018)
16. Jánošíková, Ľ: Emergency medical service planning. Commun. Sci. Lett. Univ. Žilina **9**(2), 64–68 (2007)
17. Jenelius, E.: Network structure and travel patterns: explaining the geographical disparities of road network vulnerability. J. Transp. Geogr. **17**, 234–244 (2009)
18. Kvet, M.: Computational study of radial approach to public service system design with generalized utility. In: Digital Technologies 2014: Proceedings of the 10th International IEEE Conference, pp. 198–208 (2014)
19. Kvet, M.: Advanced radial approach to resource location problems. In: Rocha, Á., Reis, L.P. (eds.) Developments and Advances in Intelligent Systems and Applications. SCI, vol. 718, pp. 29–48. Springer, Cham (2018). https://doi.org/10.1007/978-3-319-58965-7_3
20. Kvet, M.: Fast approximate algorithm for robust emergency system design. In: 20th International Carpathian Control Conference, Krakow, pp. 1–6 (2019)
21. Kvet, M.: Complexity and scenario robust service system design. In: Information and Digital Technologies 2019: Conference Proceedings, pp. 271–274 (2019)
22. Kvet, M., Janáček, J.: Hill-Climbing algorithm for robust emergency system design with return preventing constraints. Contemp. Issues Econ. **2017**, 156–165 (2017)
23. Kvet, M., Janáček, J.: Struggle with curse of dimensionality in robust emergency system design. In: Mathematical Methods in Economics 2017, pp. 396–401 (2017)
24. Kvet, M., Janáček, J.: Reengineering of the emergency service system under generalized disutility. In: 7th International Conference on Operations Research and Enterprise Systems, ICORES 2018, Madeira, Portugal, pp. 85–93 (2018)
25. Kvet, M., Janáček, J.: Robust emergency system design using reengineering approach. In: 9th International Conference on Operations Research and Enterprise Systems, ICORES 2020, Valletta, Malta, pp. 172–178 (2020)
26. Kvet, M., Janáček, J., Kvet, M.: Computational study of emergency service system reengineering under generalized disutility. In: Parlier, G.H., Liberatore, F., Demange, M. (eds.) ICORES 2018. CCIS, vol. 966, pp. 198–219. Springer, Cham (2019). https://doi.org/10.1007/978-3-030-16035-7_11
27. Majer, T., Palúch, S.: Rescue system resistance to failures in transport network. In: Proceedings of 34th International Conference Mathematical Methods in Economics, Liberec: Technical University of Liberec, pp. 518–522 (2016)
28. Marianov, V., Serra, D.: Location problems in the public sector. In: Drezner, Z. (ed.) Facility Location - Applications and Theory, pp. 119–150. Springer, Berlin (2002)
29. Ramík, J., Vlach, M.: Generalized Concavity in Fuzzy Optimization and Decision Analysis, p. 296. Kluwer Academic Publishers, Boston (2002)
30. Pan, Y., Du, Y., Wei, Z.: Reliable facility system design subject to edge failures. Am. J. Oper. Res. **4**, 164–172 (2014)

31. Reuter-Oppermann, M., van den Berg, P.L., Vile, J.L.: Logistics for emergency medical service systems. Health Syst. **6**(3), 187–208 (2017)
32. Scaparra, M.P., Church, R.L.: Location Problems under Disaster Events. Location Science, eds. Laporte, Nikel, Saldanha da Gama (2015)
33. Schneeberger, K., Doerner, K.F., Kurz, A., Schilde, M.: Ambulance location and relocation models in a crisis. CEJOR **24**(1), 1–27 (2014). https://doi.org/10.1007/s10100-014-0358-3
34. Snyder, L.V., Daskin, M.S.: Reliability models for facility location; the expected failure cost case. Transp. Sci. **39**(3), 400–416 (2005)
35. Sullivan, J.L., Novak, D.C., Aultman-Hall, L., Scott, D.M.: Identifying critical road segments and measuring system-wide robustness in transportation networks with isolating links: a link-based capacity-reduction problem. Transp. Res. Part A **44**, 323–336 (2010)
36. Teodorovič, D., Vukadinovič, K.: Traffic Control and Transport Planning: A Fuzzy Sets and Neural Networks Approach. Kluwer Academic Publishers, Boston (1998)

# Learning-Based Prediction of Conditional Wait Time Distributions in Multiskill Call Centers

Mamadou Thiongane[1], Wyean Chan[2], and Pierre L'Ecuyer[2(✉)]

[1] Department of Mathematics and Computer Science, University Cheikh Anta Diop, Dakar, Senegal
mamadou.thiongane@ucad.edu.sn
[2] DIRO, Université de Montréal, Montréal, QC, Canada
{chanwyea,lecuyer}@iro.umontreal.ca

**Abstract.** Based on data from real call centers, we develop, test, and compare forecasting methods to predict the waiting time of a call upon its arrival to the center, or more generally of a customer arriving to a service system. We are interested not only in estimating the expected waiting time, but also its probability distribution (or density), conditional on the current state of the system (e.g., the current time, queue sizes, set of agents at work, etc.). We do this in a multiskill setting, with different call types, agents with different sets of skills, and arbitrary rules for matching each call to an agent. Our approach relies on advanced regression and automatic learning techniques such as spline regression, random forests, and artificial neural networks. We also explain how we select the input variables for the predictors.

**Keywords:** Delay prediction · Wait time · Distributional forecast · Automatic learning · Service systems · Multiskill call centers

## 1 Introduction

### 1.1 How Long Will I Wait?

You make a phone call to reach your airline, bank, utility, or credit card provider, or some government service, and get a too familiar message: "All our agents are currently busy. Your call is very important to us. Please hold and one of our representatives will be with you as soon as possible." or "We currently experience a larger volume of calls than usual." Most often, the message gives no information on how much you have to wait. Sometimes, it provide a forecast of your waiting time, but the forecast can be quite inaccurate and you have no idea of its accuracy, so it may be more misleading than useful.

In a dream situation, you would be told upon arrival the exact time when the service will start. Then you could do some other activity and show up (or pick the phone) only at the right moment. Of course, this is unrealistic. The wait time is usually random and often hard to predict. A more reasonable wish could be to receive an estimate of your *expected* wait time, conditional on the current state of the system upon arrival. This conditional expectation can be computed exactly for very simple models, but in

© Springer Nature Switzerland AG 2022
G. H. Parlier et al. (Eds.): ICORES 2020/2021, CCIS 1623, pp. 83–106, 2022.
https://doi.org/10.1007/978-3-031-10725-2_5

more realistic and complex systems with multiple types of customers, different types of servers, and nontrivial routing and priority rules, it is generally hard to compute and even to approximate.

Telling the customer its conditional expected wait time upon arrival only provides limited and unsatisfactory information. If you are told "your predicted waiting time is 18 min" with no additional qualification, and you end up waiting 28 min, or you come back after 15 min only to find that you have missed your turn, you may conclude that those predictions are not so useful in the end. An improvement can be to provide a *prediction interval* (PI), such as "we predict with 95% confidence that your waiting time will be between 13 and 25 min." Or even better, a plot of the density of the waiting time distribution conditional on the current state of the system, with PIs indicated on the plot. The aim of this paper is to propose methods that can compute such *distributional forecasts* and compare their performance on real data taken from a call center.

Figure 1 gives an example of what information could be shown on the phone screen of a customer, for a call made at 13:00:00. The plot gives an estimate of the conditional density of the time at which the call will be answered, with a red line marking the expectation and two green lines indicating a 90% PI (they are at the 5% and 95% quantiles of the predicted distribution). The estimated expected answering time is at 13:05:43, and the PI is (13:03:56, 13:08:17). A key question is: How can we construct such predictors?

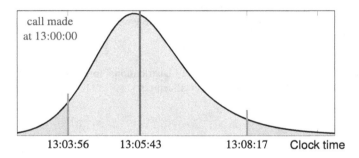

**Fig. 1.** Predicted distribution of the answering time of a call when it arrives, at 13:00:00. The red line marks the expected answering time and the two green lines are the boundaries of a 90% PI. Each tail filled in orange contains 5% of the probability density. (Color figure online)

One simple model for which the exact distribution of the wait time can be computed when the call arrives, conditional on the number of waiting calls (the queue length) at that time, is a G/M/s queue [7,17,26]. In that model, the arrival process is arbitrary, but there is a single call type, the calls are answered by order of arrival, there are $s$ identical servers, each one handling one call at a time, and the service times are assumed independent with an exponential distribution of rate $\mu$. In that case, if all servers are busy and there are $q$ other calls waiting in queue when a call arrives, then the wait time of this arriving call is the sum of $q + 1$ independent exponential random variables with rate $\mu s$. This is an Erlang random variable with shape parameter $q + 1$ and rate parameter $\mu s$. It has mean $(q + 1)/(\mu s)$, variance $(q + 1)/(\mu s)^2$, and density $f(x) = (\mu s)^{q+1} x^q e^{-\mu s x}/q!$ for $x > 0$. The explanation is that the arriving call must wait for

$q + 1$ ends of service before getting attention, and the times between the successive ends of service when the $s$ servers are busy are independent and exponential with rate $\mu s$. Independent exponential service times are rather unrealistic and, more importantly, in this paper we are interested in multiskill call centers, with different call types and separate groups of agents having different skill sets (each group can handle a different subset of the call types). No analytic formula for the wait time density or expectation is available for this case, and making the conditional predictions is much harder.

Our discussion so far was in terms of phone calls to a call center, but the models and methods studied in this paper apply more generally to "customers" or "users" who have to wait for a service; for example patients arriving to a medical clinic, or people grabbing a numbered ticket on their arrival to a store or at some governmental service such as the passport office. In the following, we use "call" and "customer" interchangeably, an "agent" is the same as a "server", and the "wait time" is also called "delay."

## 1.2   Brief Review of Earlier Work

Previous research on wait time prediction was mostly for systems with a single type of customers and identical servers. There are two main categories of proposed prediction methods for that case: the *queue-length* (QL) predictors, and the *delay-history* (DH) predictors. QL predictors use the queue length at the customer arrival, together with some system parameters, to predict the wait time. The exact analytic formula given earlier for the G/M/$s$ example is an example of a QL predictor. It can predict not only the expectation, but also the density of the wait time, and it is the best possible predictor in that situation. This type of QL predictor has been studied by [21,23,24,39]. DH predictors use the wait times of the previous customers to predict the wait time of a new arriving customer; see [2,11,22,30,36]. For example, one can predict the wait time of this customer by the wait time of the most recent customer of the same type who had to wait and already started its service (so we know its wait time), or maybe the average wait time of a few of those (e.g., the three to ten most recent ones). There are other variants. These predictors are generally designed to produce only a point estimate, i.e., only predict the expected wait time. They are further discussed in Sect. 2.1. For queueing systems with multiple customer types but a single group of identical agents that can handle all types, delay predictors based on QL and DH have been examined in [32].

For the general multiskill setting with multiple types of customers, where each type has its own queue, each server can handle only a subset of these types, and the matching between customers and servers can be done using complicated rules, predicting the delay is much more difficult. Very little has been done so far for this situation. A simple QL predictor that looks only at the queue length for the type of the arriving customer does not work well, because it neglects too much information; e.g., the lengths of the queues for the other types that the agents serving this type can also serve. If the agents that can serve the arriving type are too busy serving some other types, then the arriving customer can wait much longer. It appears difficult to extend the QL predictors to multiskill settings. For this reason, earlier work used mostly DH predictors. They are easy to apply even for complicated multiskill systems, but unfortunately they often give large prediction errors. See [35,36] and Sect. 2.1.

The basic idea of a *point predictor* for the delay is to select a set of input variables that represent the current relevant information (the state of the system), and define a real-valued *predicting function* of these variables that would represent the expected delay given the current information. This is a multivariate function approximation problem. Given available data, finding a good candidate function is a (nonlinear) multivariate regression problem. There are many ways to perform this regression and some have been examined in the literature. In particular, Thiongane et al. [35] proposed and compared such data-based delay predictors for multiskill queueing systems, using *regression splines* (RS) and *artificial neural networks* (ANN) to define (or learn) the predicting function. The input variables were the lengths of selected queues and the wait time of the most recent customer of the same type having started its service. The different proposed methods were compared on simulated models of call centers. In a similar vein, *lasso regression* (LR) was explored in [1] to predict wait times in emergency health-care units. The input variables included the queue lengths for the different priority levels, the overall load, etc. Some of these variables correspond to QL predictors that are not applicable for general multiskill call centers. Nevertheless, LR is also usable for multiskill call centers, with appropriate inputs. In this health-care application, the patients are classified by priority levels (the priority corresponds to the customer type), and the agent groups are defined differently than by subsets of the types; they may correspond to doctors, nurses, assistants, etc.

Thiongane et al. [37] made further comparisons between various types of DH predictors and regression-based (or learning-based) methods, including multilayer feedforward neural networks, using data taken from a real call center. They also proposed a method to select a set of relevant input variables. The performance of these regression-type predictors depend very much on which input variables are considered, and on how much relevant data is available for the learning. If important variables are left out, the performance may degrade significantly. If there is too little data, or if the current setting differs too much from the one in which the data was obtained, then the forecasting error is also likely to be large. Too little data combined with too many input variables also lead to overfitting.

All the methods and papers discussed so far for the multiskill systems are for *point predictions* only, i.e., to predict the conditional expected wait time (a single number), and not to predict its distribution. A few papers examine quantile prediction of the waiting times of patients in health-care emergency departments [10, 34]. The authors predict the median, 10%, 90% or 95% percentiles of the wait times using multiple regression.

### 1.3   Contribution and Outline

The present paper is a follow up to [37]. As an important extension, we propose methods to estimate the *conditional density* and *quantiles* of the wait time, as in Fig. 1, instead of only the expectation. We also report additional experiments, and use data from a second call center.

The rest of the paper is organized as follows. In the next section, we specify and discuss the DH and regression-based delay predictors considered in our experiments. In Sect. 3, we explain how we propose to estimate the conditional density of the wait time. In Sect. 4, we describe the experiment setup, and in Sect. 5 and 6, we report on numerical experiments with data from two different call centers. The first one is the call center of an Israeli bank, and the second one is from an information technology (IT) company in the Netherlands. For each one, we describe the data, and we compare our different point predictors and density predictors for the delay. Section 7 provides a conclusion.

## 2  Point Predictors for Multiskill Systems

We now discuss different types of point predictors used and compared in our experiments. These predictors return a single number, which may be interpreted as an estimate of the expected delay. They are all "learning-based" in some sense, although for the DH predictors, the learning is based only on the delays of the very recent customers. We exclude QL predictors, since they are not adapted to multiskill settings.

### 2.1  Delay-History Predictors

In a multiskill call center, a DH predictor estimates the wait time of a new arrival by looking only at the delays of the most recent customers of the same type who started their service. There is no learning based on lots of data to estimate function parameters, so these predictors are simple and easy to implement. The DH predictors discussed here are the best performers according to our previous experiments in [35–37].

The simplest and most popular DH predictor is the *Last-to-Enter-Service* (LES) predictor. It returns the wait time of the most recent customer of the same type among those who had to wait and have started their service [22].

A generalization often used in practice [11] is to take the $N$ most recent customers of the same type who had to wait and have started their service, for some fixed positive integer $N$, and average their wait times. This is the *Averaged LES* (Avg-LES).

One variant of the Avg-LES also takes the average of the wait times of past customers of the same type who had to wait and have started their service, but the average is only over the customers who found the same queue length as the current one when they arrived. This predictor was introduced in [36] and was the best performing DH predictor in the experiments made in that paper. It is called the *Average LES Conditional on Queue Length* (AvgC-LES).

Another one is the *Extrapolated LES* (E-LES), defined as follows [36]. For a new arriving customer, it looks at the delay information of all customers of the same type that are *currently waiting in queue*. The final delays of these customers are still unknown, but the (partial) delays elapsed so far are extrapolated to predict the final wait times of these customers. E-LES returns a weighted average of these extrapolated delays, as explained in [36].

The *Proportional Queue LES* (P-LES) predictor starts with the delay $d$ of the LES customer and makes an adjustment to account for the difference in the queue length $q_{\text{LES}}$ at the arrival of this LES and the current queue length $q$ when the new customer arrives [20]. The adjusted predictor is

$$D = d(q + 1)/(q_{\text{LES}} + 1).$$

## 2.2   Regression-Based Predictors

A regression-based predictor constructs a multivariate *predictor function* of selected input variables deemed important. It approximates the conditional expectation of the delay $W$ of an arriving customer of type $k$, conditional on the current state of the system, which is represented by the vector $\mathbf{x}$ of these input variables. The predictor function for customer type $k$ is $p_{k,\theta(k)}$, where $\theta(k)$ is a vector of parameters which depends on $k$. It must be estimated (learned) from the data in a training step. The predicted delay when in state $\mathbf{x}$ will be $p_{k,\theta(k)}(\mathbf{x})$. Constructing this type of predictor involves three main parts that are inter-related, for each $k$: (a) selecting which variables to put in $\mathbf{x}$; (b) selecting the general form of $p_{k,\theta(k)}$; and (c) estimating (learning) the parameter vector $\theta(k)$. In the remainder of this section, we explain how we have implemented these three parts in our experiments. The method used for part (c) depends very much on the choice of predictor function in part (b). For that, we will consider and compare the following three choices: (1) a smoothing (regression) cubic spline additive in the input variables (RS), (2) a lasso (linear) regression (LR), and (3) a deep feedforward multilayer artificial neural network (ANN).

**Identifying the Important Variables.** In the G/M/$s$ queuing system discussed earlier, the analytic formula tells us clearly that the only important input variables are the number $s$ of servers and the number $q$ of customers in the queue. Everything else is irrelevant. For more complex multiskill systems, however, identifying the most relevant inputs for the prediction for a given customer type $k$ is not so simple. Leaving out important variables is bad, and keeping too many of them leads to overfitting, as is well-known from regression theory.

For our numerical examples with the call center data, we used the following methodology. We started by putting as candidates all the observable variables that could have a chance of helping the prediction, then we used a feature selection algorithm to perform a screening among them. The candidate input variables were the following: the vector $\mathbf{q}$ of current queue lengths for all call types; the number $s$ of agents that are serving the given call type, the total number $n$ of agents currently working in the system, the current time $t$ (when the call arrives), the wait time of the $N$ most recently served customers of the given call type, and the delay predicted by the DH predictors LES, P-LES, E-LES, Avg-LES, and AvgC-LES.

To screen out and make a selection among these inputs, we used a technique based on the *random forest* (RF) bootstrapping methodology of Breiman [6]. Among the various feature selection algorithms based on this methodology, we picked *Boruta* [27], which was the best performer in empirical comparisons between many selection algorithms in [9]. The general idea of random forests is to generate a forest of decision trees whose nodes correspond to the selection decisions for the input variables. Boruta extends the data by adding copies of all input variables, and reshuffles these variables to reduce their correlations with the response. These reshuffled copies are named the *shadow features*. Boruta then runs a random forest classifier on this extended data set. It makes bootstrap samples on the training set and constructs decision trees from these samples. The importance for each input variable is assessed by measuring the loss of accuracy of the model when the values of this input are permuted randomly across the observations. This measure is called the *mean decrease accuracy*. It is computed separately for all trees of the forest that use the given input variable. The average and standard deviation of the loss of accuracy is then computed for each input, a $Z$ score is obtained by dividing the average loss by its standard deviation, and this score is used as the importance measure. The maximum $Z$-score among the shadow features (MZSA) is used to select the variables deemed useful to predict the delay. The input variables are then ranked according to these scores, and those with the highest scores are selected. The variables whose $Z$-scores are significantly lower than MZSA are declared "unimportant", those whose $Z$-scores are significantly higher than MZSA are declared "important" [27], and decisions about the other ones are made using other rules.

**Measuring the Prediction Error.** The parameter vector $\theta$ is estimated by minimizing the *mean squared error* (MSE) of point predictions. That is, if $E = p_{k,\theta(k)}(\mathbf{x})$ is the predicted delay for a given customer of type $k$ who receives service after some realized wait time $W$, the MSE for type $k$ calls is

$$\mathrm{MSE}_k = \mathbb{E}[(W - E)^2].$$

This expectation cannot be computed exactly, but we can estimate it by its empirical counterpart, the *average squared error* (ASE), defined as

$$\mathrm{ASE}_k = \frac{1}{C_k} \sum_{c=1}^{C_k} (W_{k,c} - E_{k,c})^2 \tag{1}$$

for customer type $k$, where $C_k$ is the number of customers of type $k$ who had to wait and for which we made a prediction. In the end, we use a normalized version of the ASE, called the *root relative average squared error* (RRASE), defined as the square root of the ASE divided by the average wait time of the $C_k$ served customers, rescaled by a factor of 100:

$$\mathrm{RRASE}_k = \frac{100 \, (\mathrm{ASE}_k)^{1/2}}{(1/C_k) \sum_{c=1}^{C_k} W_{k,c}}.$$

We estimate the parameter vector $\theta(k)$ for each $k$ in this way from a learning data set that represent 80% of the collected data. The other 20% of the data is saved to measure and compare the accuracy of these delay predictors.

**Regression Splines (RS).** *Regression splines* (RS) are a powerful class of approximation methods for general smooth functions [8,25,40]. Here we use smoothing additive cubic splines, for which the parameters are estimated by least-squares regression after adding a penalty term on the function variation to favor more smoothness. If the information vector is written as $\mathbf{x} = (x_1, \ldots, x_D)$, the additive spline predictor can be written as

$$p_{k,\theta(k)}(\mathbf{x}) = \sum_{d=1}^{D} f_d(x_d),$$

where each $f_d$ is a one dimensional cubic spline. The parameters of all these spline functions $f_d$ form the vector $\theta$. We estimated these parameters using the function gam from the R package mgcv [41].

**Lasso Regression (LR).** Lasso Regression is a type of linear regression [13,25,38] with a penalty term proportional to the sum of absolute values of the magnitude of coefficients, added to reduce overfitting. The LR predictor is

$$p_{k,\theta(k)}(\mathbf{x}) = \beta_0 + \sum_{d=1}^{D} \beta_d \cdot x_d,$$

where $\mathbf{x} = (x_1, \ldots, x_D)$ is the input vector, as in ordinary linear regression, but the vector of coefficients $\theta(k) = (\beta_0, \beta_1, \ldots, \beta_D)$ is selected to minimize the sum of squares of errors plus the penalty term. To estimate the parameter vector $\theta(k)$, we used the function glmnet in the R package gmlnet [12].

**Artificial Neural Networks (ANN).** An *artificial neural network* (ANN) is an effective tool to approximate complicated high-dimensional functions [4,28]. Here we use a deep feedforward ANN, which contains one input layer, one output layer, and a few intermediate (hidden) layers. The outputs from the nodes at layer $\ell$ are the inputs for all nodes at layer $\ell + 1$. Each node of the input layer corresponds to one element of the input vector $\mathbf{x}$. The output layer has a single node, which returns the predicted delay. At each hidden node, we have a rectifier activation function of the general form $h(\mathbf{z}) = \max(0, b + \mathbf{w} \cdot \mathbf{z})$, where $\mathbf{z}$ is the input vector for this node, whereas $b$ and the vector $\mathbf{w}$ are parameters learned by training [14]. At the output node, to predict the delay, we use a linear activation function of the form $h(\mathbf{z}) = b + \mathbf{w} \cdot \mathbf{z}$ where $\mathbf{z}$ is the vector of outputs from the nodes at the previous hidden layer. Here, the vector $\theta$ represents the set of all the parameters $b$ and $\mathbf{w}$, over all the nodes of the ANN. Good parameter values are learned by a back-propagation algorithm that relies on a stochastic gradient descent method. Several hyperparameters used in the training are determined empirically. For a guide on training, see [3,5,15,18]. For this paper, we did the training using the Pylearn2 software [16].

## 3   Density Predictors

We saw that for a G/M/$s$ queue, the delay of a customer who finds $q$ waiting customers in front of him upon arrival has an exact Erlang distribution whose density has an explicit formula. But for more complex multiskill systems, the distribution of the delay conditional on the current state of the system has an unknown density which is likely to be very complicated and is much harder to estimate. Estimating a general univariate density in a non-parametric way, from a given data set coming from this density, is already a difficult problem in statistics. With the best available methods, e.g., kernel density estimators, the error in the density estimate converges at a slower rate than the canonical rate of $\mathcal{O}(n^{-1/2})$ as a function of the number $n$ of observations [31]. Estimating a conditional density which is a multivariate function of several input variables is even more difficult.

After a few initial attempts, we decided not to estimate directly the density of the delay time conditional on $\mathbf{x}$, but we consider alternatives that fit the prediction errors to a simpler parametric model conditional on selected information from $\mathbf{x}$, and kernel density estimators that depend on this limited information. We take our point predictor of the delay (the estimate of the expected wait time), and add to it the estimated density of the prediction error. That is, for each predictor and each call type, we first estimate (or learn) the parameters of the point predictor, then compute the prediction error for each customer. After that, we fit a parametric or non-parametric density to these errors, or we train a learning algorithm to model these errors conditional on $\mathbf{x}$. This provides an estimate of the density of the prediction error. By adding the point predictor to this density, we obtain a predicted density for the delay.

Specifically, let us write the wait time $W$ of a customer as

$$W = p_{k,\theta(k)}(\mathbf{x}) + \epsilon(k, \mathbf{x}), \qquad (2)$$

where $\epsilon(k, \mathbf{x})$ is the prediction error. The idea is to estimate the density of $\epsilon(k, \mathbf{x})$ and this will give us the density of $W$, since $p_{k,\theta(k)}(\mathbf{x})$ is a constant. The density of $\epsilon(k, \mathbf{x})$ certainly depends on $k$ and $\mathbf{x}$, so a key issue is how to model this dependence. A very simple solution would be to ignore the dependence on $\mathbf{x}$ and just pick a density for each $k$. A more refined solution is to partition the space of values of $\mathbf{x}$ in a small number of subsets and estimate one density for each subset. The number of subsets should not be too large, because it would lead to estimating too many different densities, and eventually to overfitting.

In this paper, we consider two different ways to partition the domain space of $\mathbf{x}$. The first approach uses clearly defined cuts of the space defined by selecting which *features* of $\mathbf{x}$ to emphasize and which ones are to be aggregated. The second method uses a learning-based algorithm, in which the partitions are not explicitly defined. Note that these approaches can be used to estimate the error density of any predictor, and are not limited to the RS, LR, and ANN methods presented in Sect. 2.

## 3.1  Parametric Model Conditional on Queue Length

To predict the mean wait time, we trained a function of the entire vector $\mathbf{x}$. But predicting an entire density as a function of $\mathbf{x}$ is more difficult. With the approach described here, we will make our prediction as a function of a more limited amount of information. We partition the space of $\mathbf{x}$ according to the most important predictive features for the waiting time, and aggregate the rest of the features. Suppose we are predicting the density for call type $k$. For our case studies, the Boruta algorithm described in Sect. 2.2 selected the queue length $q_k$ of call type $k$ as the most important feature. This motivates the idea of fitting a parametric probability model of the prediction errors $\epsilon(k, \mathbf{x})$ conditional only to the queue length $q_k$. That is, we would fit a different model for each value of $q_k$. By doing that, a lot of information in $\mathbf{x}$ is ignored, but we hope that much of the variability on $\epsilon(k, \mathbf{x})$ is captured by $q_k$.

In our numerical studies, we tried to fit several distributions to the realizations of $\epsilon(k, \mathbf{x})$ conditional on $q_k$, including the gamma, lognormal, log-logistic, and Weibull distributions, and shifted version of them. Shifting these distributions is necessary because $\epsilon(k, \mathbf{x})$ must be allowed to take negative values. That is, before fitting one of the distributions named above, we first shift the errors by adding a positive constant $\tau_k(q)$ large enough so that $\tau_k(q) + \epsilon(k, \mathbf{x}) > 0$ for all $\mathbf{x}$ with $q_k = q$. Then the distribution is fitted to the shifted observations. After that, a negative shift of $-\tau_k(i)$ is applied to obtain the distribution of the prediction error. For each call type $k$ and queue length $q_k$, we used a grid search to find the best value for $\tau_k(q_k)$. In our case studies, the *shifted log-logistic* gave the best fit in most cases, and the best shift did not depend much on $q_k$. One explanation for this could be that both the mean and standard deviation of the delay increase approximately linearly in $q_k$, so the starting point of the distribution does not change much. As an illustration, Fig. 2 compares the density functions of the errors fitted with the log-logistic, gamma, lognormal, and Weibull distributions, for some call type $k$ in our data, conditional to $q_k = 2$. For this figure and for all the numerical results reported in the paper, we fitted the parametric models using the R package `fitdistrplus`, and the package `actuar` for the log-logistic distribution. For comparison, the figure also shows a *kernel density estimator* (KDE) conditional on type $k$ and $q_k = 2$. This KDE was obtained as discussed in Sect. 3.2.

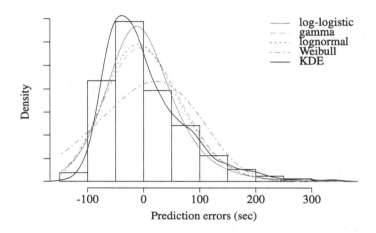

**Fig. 2.** The distribution of the prediction error made by the ANN predictor, conditional to $q_k = 2$, for some call type $k$, and four parametric densities (log-logistic, gamma, lognormal, and Weibull) fitted to this data. The black curve is a KDE. Among the four parametric densities, the log-logistic gives the best fit.

For each $k$ and each value of $q_k$ until a certain threshold where the amount of training data becomes too small, we fit a different shifted log-logistic density. For queue lengths that have insufficient training data, we can pool together several values of $q_k$ and fit one density for them. In the numerical section, we set the threshold to $q_k = 5$, and all states $\mathbf{x}$ for which $q_k > 5$ are pooled in a single group, for each $k$. Interestingly, when observing the empirical mean and standard deviation of the prediction error conditional to $q_k$, we find that we can also fit a simple model for the parameters of the log-logistic as a function of $q_k$. By design, the mean of the shifted log-logistic density for the prediction error should be zero (or near zero). We have observed that the standard deviation is not far from affine in $q_k$. Figure 3 gives an illustration. This means that we can fit a linear regression model for the standard deviation as a function of $q_k$, for each $k$. The scale and shape parameters of the log-logistic distribution can then be determined by the mean and standard deviation.

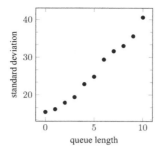

**Fig. 3.** Standard deviation (in seconds) of the prediction error made by the ANN predictor conditional on the queue length, for one call type in the test data set.

According to Boruta, the second most important variable for the prediction after $q_k$ is $t$, the arrival time of the call during the day. Therefore, to further refine the partition of the space of values of $\mathbf{x}$, we could model the density of the prediction error as a function of both $q_k$ and $t$. We did not do that for the numerical experiments reported here, but this could be explored in the future.

### 3.2   A Non Parametric Option: Using a Kernel Density Estimator

Parametric distributions are attractive because the entire density is defined by just a few parameters (the scale, shape, and location parameters for the shifted log-logistic). However, the true density function of the prediction error may have a shape that hardly matches any of the common families of distributions. A *kernel density estimator* (KDE) provides a much more flexible non-parametric solution. The KDE can be used directly to estimate the density of the prediction error. Figure 4 compares the KDE with the best fitted shifted log-logistic. We see a significant gap between the two densities for small queue length $q_k$. For this figure and for all the numerical results reported in this paper, the KDE was computed using the Epanechnikov kernel and the bandwidth was selected using the heuristic of Silverman [19, 33].

One drawback of the KDE is that it performs poorly when there are too few observations, and it is also difficult to extrapolate the density obtained from the KDE for frequent values of $q_k$ to rare (larger) values, as we did with the parametric log-logistic density. For this reason, we may want to use a combined model: construct the KDE for queue lengths $q_k$ that have a large number of observations, and use a parametric model to extrapolate the density for less frequent data points. Another possibility could be to use the KDE with a scaling factor proportional to $q_k$, but we did not do that.

### 3.3   Quantile Regression with Random Forest

In the two previous subsections, we partitioned the space of values of $\mathbf{x}$ manually by reducing the number of dimensions on $\mathbf{x}$, and fitted a different distribution over each piece of the partition, for each $k$. This makes a lot of distributions and it can hardly be used unless we select only a few important features from $\mathbf{x}$, say one or two. In this section, we take a different approach. We do not partition explicitly the space of values

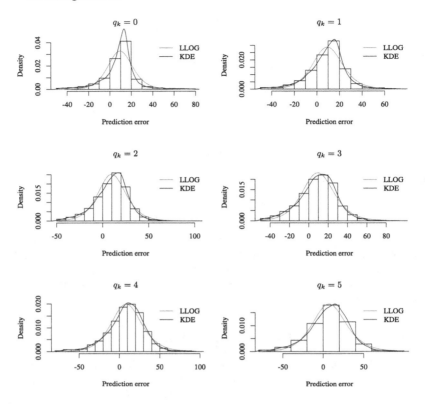

**Fig. 4.** Log-logistic density (red) and KDE (black) for the prediction error (in seconds) made by an ANN predictor, conditional to queue lengths $q_k$ from 0 to 5. (Color figure online)

of $\mathbf{x}$, but we rather leave this task to a learning-based algorithm. More specifically, we use a *quantile random forest* (QRF) algorithm that takes as input the entire vector $\mathbf{x}$, trains an ensemble of decision trees, and outputs the desired quantile and density functions. The partition rules learned by QRF can be very complex, much more than simply aggregating $\mathbf{x}$ conditionally to the queue length as in Sect. 3.1.

The QRF is implemented and trained identically to an ordinary RF, except that each leaf of a tree retains in memory the list of errors $\epsilon(k, \mathbf{x})$ for all training predictors $\mathbf{x}$ that ended in this leaf, instead of only the average $\epsilon(k, \mathbf{x})$ as in a RF. That is, QRF is not trained specifically to predict the quantiles or the density function, but it is a by-product of an RF trained to predict the expectation of $\epsilon(k, \mathbf{x})$. Each decision tree $v$ of the QRF defines a complete partition of the space of $\mathbf{x}$, and consequently of the training data set, say $\hat{\mathbf{X}} = \cup_{l=1}^{L_v} \hat{\mathbf{X}}_l^v$, where $L_v$ is the number of leaves in tree $v$. Identical values of $\mathbf{x}$ always fall in the same leaf of the tree $v$. However a leaf can be the end node for a large number of different states $\mathbf{x}$, with different prediction errors $\epsilon(k, \mathbf{x})$. The collection of these $\epsilon(k, \mathbf{x})$ is then used to estimate the probability distribution of $\epsilon(k, \mathbf{x})$ for the $\mathbf{x}$ that belong to that leaf.

After completing the training stage, QRF estimates the $\alpha$-th quantile of the error $\epsilon(k, \mathbf{x}')$ for a given input $\mathbf{x}'$ as follows. For each decision tree $v$, it identifies the terminal leaf $l_v(\mathbf{x}')$ to which $\mathbf{x}'$ belongs, after traversing through the branches of the tree. Next, it computes the empirical $\alpha$-th quantile from the saved list of all values of $\epsilon(k, \mathbf{x})$ for $\mathbf{x} \in \hat{\mathbf{X}}^v_{l_v(\mathbf{x}')}$ in the training set. These steps are repeated for every decision tree in QRF, and a weighted average of the $\alpha$-th quantiles is finally returned.

To estimate the density conditional on some state $\mathbf{x}'$ with the QRF, we take the saved list of all values of $\epsilon(k, \mathbf{x})$ for $\mathbf{x} \in \hat{\mathbf{X}}^v_{l_v(\mathbf{x}')}$ in the training set, and merge these lists for all trees $v$ into a single list of observations. Then a KDE is constructed by using these observations, to obtain a predicted density that depends on $\mathbf{x}'$. Figure 5 shows an example with five density functions estimated via QRF (in black) for five different vectors $\mathbf{x}$ with the same queue length $q_k = 2$. These densities are compared with the direct KDEs (in red) obtained as explained in Sect. 3.2, by using all the values of $\mathbf{x}$ for which $q_k = 2$. The QRF-based density estimator depends on more than just $k$ and $q_k$.

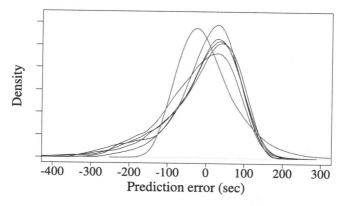

**Fig. 5.** Densities predicted by QRF (in black) for five different input vectors $\mathbf{x}$, lying in different tree leaves but having the same queue size $q_k = 2$, for some call type $k$. The KDEs obtained for all $\mathbf{x}$ with $q_k = 2$ are shown in red, for comparison. (Color figure online)

Training a QRF model requires substantially more computing power and CPU time than fitting a parametric distribution. It also involves many more parameters, which need to be saved after the training. On the other hand, we only need to train one QRF for each call type, since the quantile regression is conditional on the entire vector $\mathbf{x}$. QRF has likely a higher learning capacity than a parametric model, but it also faces a larger risk of overfitting. In Sect. 5, we compare it empirically to the more classical KDE and to parametric methods, with the call center data. In all our experiments, we use the QRF implemented in the R package `quantregForest` [29]. The learning capacity can be adjusted by changing the number and the depth of the decision trees, but the results reported in this paper were obtained with the default values.

## 4   Numerical Experiment Setup

Before presenting our numerical results, we describe our methodology for comparing the prediction algorithms for the expected delay times and the density functions. We have data for two call centers. The first is the call center of a small bank in Israel and the second is from a large information technology (IT) company located in the Netherlands. We will refer to them as the Bank and IT call centers, respectively. The data set for each call center contains exactly one year of observations, from January 1 to December 31 (but for different year). We partition each data set chronologically as follows: the first 80% of the data is the *training set* and the remaining 20% is the *test set*.

First, we identify the important predictive features of the delay time by using the Boruta algorithm, as explained in Sect. 2.2. For both the Bank and IT call centers, the features vector was chosen as $\mathbf{x} = (\mathbf{q}, a, l, s, n, t)$ where $\mathbf{q}$ is the vector of queue sizes for all call types, $a$ and $l$ are the delay prediction by AvgC-LES and by LES, respectively, $s$ is the number of agents that can serve this call, $n$ is the total number of agents, and $t$ is the time of arrival of the call.

We train the regression-based predictors (RS, LR, ANN) to predict the mean waiting time, using the training set. We compare their accuracy to those of the (simpler) DH predictors, which do not require any training, based on the RRASE score, computed over the test set.

To compare the density predictors, we select ANN as the point predictor of the mean delay time, because it often displays the best accuracy in our experiments. The training set for the density predictors is the set of prediction errors obtained by the ANN over its training set. Then, we fit or train a parametric log-logistic distribution (LLOG), a KDE, and a QRF.

Because the true density function of the prediction error is unknown, we compare the coverages of 90% PIs instead, as well as the coverages of the tails. That is, we use the estimated conditional density to compute a 90% PI on the answering time (or equivalently the wait time) of each call. Once we know the true answering time, we can check where it lies among the three following possibilities (1) to the left of the PI; (2) inside the PI; (3) to the right of the PI. Then we can compute the proportion of calls falling in each of the three categories. Ideally, there should be 5% in each of categories (1) and (3), and 90% in category (2). In our numerical experiments, we compare the observed proportions to these ideal ones. We also do the same with a 80% PI.

## 5   Experiments with Data from a Bank Call Center

### 5.1   The Call Center and Available Data

Our first data set is from the small call center of a bank in Israel, recorded over the entire year of 1999. This center operates from 7:00 to midnight on weekdays (Sunday to Thursday), and on weekends (Friday to Saturday) it closes at 14:00 on Friday and reopens at around 20:00 on Saturday. There are five inbound call types, one outbound type, and the center has eight working agents on average.

About 65% of the calls are served by the Interactive Voice Response (IVR) unit and leave without interacting with an agent. For the other 35%, the customer wants to speak to an agent. If no idle agent is available, the customer is placed in a queue and receives information on the queue size and the waiting time of the customer at the head of queue (the HOL predictor). The customers are served in first-come-first-served (FCFS) order. Table 1 gives a statistical summary of the arrival counts and wait times during the year. Type 1 has by far the largest volume.

**Table 1.** Statistical summary of arrival counts and waits for the Bank call center.

| Call type | 1 | 2 | 3 | 4 | 5 |
|---|---|---|---|---|---|
| Total number calls | 302 522 | 67 728 | 39 342 | 20 732 | 12 295 |
| Served, no wait | 42% | 34% | 36% | 30% | 44% |
| Served, waited | 46% | 36% | 56% | 51% | 46% |
| Abandon | 11% | 30% | 8% | 19% | 10% |
| Avg wait time (sec) | 99 | 145 | 121 | 167 | 138 |
| Avg service time (sec) | 187 | 124 | 263 | 369 | 274 |
| Avg queue length | 6.3 | 2.5 | 2.0 | 1.7 | 0.9 |

## 5.2  Experimental Results on Predictions

We first report in Table 2 the RRASE for six predictors of the mean delay. We see that all the learning-based predictors are more accurate than the DH predictors. RS and ANN are the most accurate. Among the DH predictors, AvgC-LES gives the best performance and is not too far from RS and ANN for call type 1.

**Table 2.** The Bank call center: RRASE of the four largest-volume call types (lower is better). The RS and ANN are the most accurate predictors here.

| Type | DH predictors | | | Learning-based predictors | | |
|---|---|---|---|---|---|---|
| | Avg-LES | LES | AvgC-LES | RS | LR | ANN |
| 1 | 0.925 | 0.941 | 0.765 | 0.731 | 0.737 | **0.701** |
| 2 | 0.980 | 0.990 | 0.933 | 0.762 | 0.849 | **0.751** |
| 3 | 0.922 | 0.941 | 0.853 | **0.725** | 0.750 | 0.737 |
| 4 | 1.348 | 1.541 | 1.320 | **1.178** | 1.205 | 1.185 |

Tables 3 and 4 compare the PI coverages for the error on the delay prediction for call type 1 by ANN, when the density of the prediction error is estimated by the QRF, the fitted LLOG, and the KDE, as explained in Sect. 3. We tried both 90% and 80% PIs. Recall that QRF is trained only once with all the vectors $\mathbf{x}$ in the training data set, whereas LLOG and KDE are fitted separately for each queue size $q_k$.

The coverages were computed as explained in the last paragraph of Sect. 4. For each call (represented by $\mathbf{x}$) in the test set, the expected delay was estimated using the ANN predictor, and the estimated 5%, 10%, 90%, and 95% quantiles of the prediction error were also computed using each of the three methods: QRF, LLOG, and KDE. Because there are few data for large queue size, all $\mathbf{x}$ for which $q_1 > 5$ were grouped together. We computed the percentages of calls falling in each of the three categories (1) to (3), for both the 90% and the 80% PI's. The percentages are reported in the tables. For call type 1, QRF clearly has a much better coverage accuracy than LLOG and KDE, with values that are fairly close to the desired coverage levels.

**Table 3.** The Bank call center, call type 1: Coverage percentage of a 90% PI and of the 5% tail on each side for the test data set, for different queue sizes.

| Queue size | <5% | | | [5%, 95%] | | | >95% | | | Total obs |
|---|---|---|---|---|---|---|---|---|---|---|
| | QRF | LLOG | KDE | QRF | LLOG | KDE | QRF | LLOG | KDE | |
| 0 | 5.05 | 10.25 | 2.76 | 92.02 | 89.21 | 96.88 | 2.94 | 0.53 | 0.35 | 6222 |
| 1 | 5.06 | 11.06 | 5.33 | 90.27 | 88.58 | 93.62 | 4.66 | 0.35 | 1.03 | 4738 |
| 2 | 5.10 | 9.42 | 13.36 | 89.28 | 90.26 | 85.97 | 5.60 | 0.31 | 0.66 | 3173 |
| 3 | 5.57 | 9.91 | 11.29 | 89.08 | 89.41 | 87.62 | 5.33 | 0.66 | 1.08 | 2117 |
| 4 | 4.47 | 8.95 | 9.84 | 91.34 | 90.00 | 87.99 | 4.17 | 1.04 | 2.16 | 1340 |
| 5 | 4.58 | 9.27 | 9.05 | 89.60 | 89.38 | 87.70 | 5.81 | 1.34 | 3.24 | 895 |
| $\geq 6$ | 5.64 | 13.14 | 10.26 | 89.95 | 85.24 | 87.71 | 5.41 | 1.62 | 2.01 | 1294 |

**Table 4.** The Bank call center, call type 1: Coverage percentage of a 80% PI and of the 10% tail on each side for the test data set, for different queue sizes.

| Queue size | <10% | | | [10%, 90%] | | | >90% | | | Total obs |
|---|---|---|---|---|---|---|---|---|---|---|
| | QRF | LLOG | KDE | QRF | LLOG | KDE | QRF | LLOG | KDE | |
| 0 | 9.77 | 14.15 | 17.76 | 81.28 | 82.29 | 81.71 | 8.95 | 3.55 | 0.53 | 6222 |
| 1 | 10.13 | 16.36 | 18.80 | 78.75 | 81.17 | 80.77 | 11.12 | 2.47 | 0.42 | 4738 |
| 2 | 9.90 | 15,73 | 16.75 | 79.55 | 81.31 | 81.54 | 10.56 | 2.96 | 1.71 | 3173 |
| 3 | 9.49 | 15.54 | 15.78 | 79.88 | 82.38 | 82.05 | 10.63 | 2.08 | 2.17 | 2117 |
| 4 | 7.90 | 14.62 | 14.54 | 83.00 | 82.48 | 81.73 | 9.10 | 2.91 | 3.73 | 1340 |
| 5 | 8.04 | 16.42 | 15.53 | 81.01 | 81.56 | 80.34 | 10.95 | 2.01 | 4.13 | 895 |
| $\geq 6$ | 10.20 | 17.77 | 15.77 | 80.14 | 80.76 | 78.83 | 9.66 | 1.47 | 5.41 | 1294 |

Table 5 shows some details on the values of the 10% and 90% quantiles that delimit a 80% PI that would be announced to a customer in the test set, aggregated by the queue size. For a fixed queue size, LLOG and KDE return constant estimates for the quantiles, because they estimate the density as a function of the queue size only. QRF, on the other hand, returns quantile estimates that depend on $\mathbf{x}$, so it can return very different values for the same queue length $q_k$. In the table, we give the mean and standard deviation of these values for each $q_k$. As expected, the width of the 80% PI increases with the queue size for all three predictors. Figure 6 compares the density functions obtained by LLOG fitting and by a KDE. LLOG has better fit when the queue is longer, but it has significant fitting error when the queue is short (sizes 0, 1 and 2).

## 6    Experiments with Data from an IT Call Center

### 6.1    The Available Data

This is a call center of an information technology (IT) company located in The Netherlands. The data was collected over the entire year of 2014, and contains a total of 1,543,164 call logs. The center operated from 8:00 to 20:00 on weekdays (Monday

**Table 5.** The Bank call center, call type 1: the mean and standard deviation of the 10% and 90% quantiles on the predictor error for QRF, and the actual quantiles for LLOG and KDE, conditional on the queue size.

| Queue size | 10% | | | | 90% | | | |
|---|---|---|---|---|---|---|---|---|
| | QRF | | LLOG | KDE | QRF | | LLOG | KDE |
| | Mean | Std dev. | | | Mean | Std dev. | | |
| 0 | −64.85 | 27.97 | −40.70 | −30.62 | 31.89 | 7.92 | 39.99 | 55.73 |
| 1 | −92.59 | 34.26 | −58.26 | −49.17 | 49.63 | 11.73 | 67.32 | 88.67 |
| 2 | −109.12 | 36.57 | −75.14 | −71.16 | 69.72 | 13.93 | 90.13 | 101.84 |
| 3 | −126.86 | 42.03 | −85.79 | −84.67 | 85.92 | 18.63 | 120.13 | 118.55 |
| 4 | −140.02 | 44.92 | −90.10 | −91.48 | 98.12 | 22.80 | 126.39 | 120.07 |
| 5 | −152.08 | 46.95 | −97.73 | −101.33 | 105.77 | 28.96 | 152.54 | 129.03 |
| ≥6 | −176.40 | 62.09 | −107.98 | −118.73 | 121.04 | 36.89 | 214.05 | 161.03 |

to Friday), and served 27 call types with approximately 312 agents. About 56% of the calls have received service immediately, 38% have waited before getting a response, and 6% have abandoned. In this study, we consider only the five call types (type 1 to 5) that represent the largest volume of incoming calls (more than 90% of the total volume). Table 6 gives a statistical summary of the arrival counts, wait times, and service times, for the five call types.

**Table 6.** The IT call center: a statistical summary of arrivals counts and waits during the year. Adapted from [37].

| | Type 1 | Type 2 | Type 3 | Type 4 | Type 5 |
|---|---|---|---|---|---|
| Total number calls | 568 554 | 270 675 | 311 523 | 112 711 | 25 839 |
| Served, no wait | 61% | 52% | 55% | 45% | 34% |
| Served, waited | 35% | 40% | 40% | 46% | 54% |
| Abandon | 4% | 7% | 5% | 8% | 12% |
| Avg wait time (sec) | 77 | 91 | 83 | 85 | 110 |
| Avg service time (sec) | 350 | 308 | 281 | 411 | 311 |
| Avg queue length | 8.2 | 3.3 | 4.4 | 4.3 | 0.9 |

We partition the weekdays in two categories, according to the arrival patterns and volumes. Monday forms its own category, while the four other weekdays (Tuesday to Friday) form the second category. For each call type $k$, we built two sets of prediction functions, one for each category. In this paper, we report prediction results for the second category, for which we have more data. Results on the prediction of the expected wait time were already presented in [37], but that paper did not consider quantile and density prediction. We summarize the wait time prediction results here for the sake of completeness.

## 6.2   Experimental Results on Predictions

Table 7 compares the RRASE values of six different predictors of the mean delay time, namely three DH predictors and three learning-based predictors. ANN is the clear winner among those six.

Tables 8 and 9 compare the PI coverages exactly as in Tables 3 and 4, for call type 1 of the IT call center. In contrast with the Bank call center, here the KDE gives the coverage closest to the target for the 95% and 90% quantiles (the right tail). For the 5% and 10% quantiles (the left tail), QRF does better when the queue size is small whereas the KDE is more accurate for longer queue sizes. LLOG lags behind.

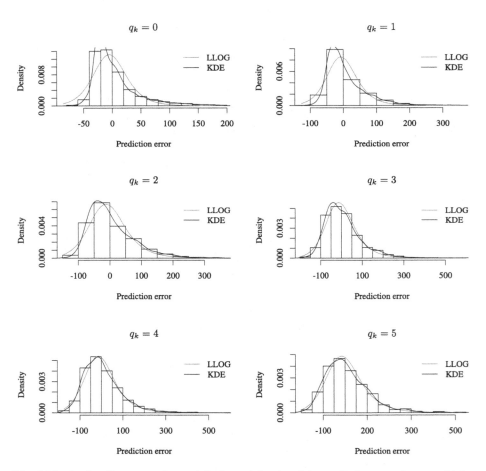

**Fig. 6.** The Bank call center, call type 1: Estimated density of the prediction error (in seconds) for fitted LLOG and the KDE, conditional on the queue size.

**Table 7.** The IT call center: the RRASE for the five call types. The ANN has the best accuracy, followed by RS.

| Type | DH predictors | | | Learning-based predictors | | |
|---|---|---|---|---|---|---|
| | Avg.LES | LES | AvgC-LES | RS | RL | ANN |
| Type 1 | 0.489 | 0.443 | 0.443 | 0.396 | 0.415 | **0.361** |
| Type 2 | 0.610 | 0.565 | 0.577 | 0.492 | 0.515 | **0.462** |
| Type 3 | 0.567 | 0.516 | 0.518 | 0.455 | 0.471 | **0.448** |
| Type 4 | 0.487 | 0.424 | 0.445 | 0.395 | 0.385 | **0.377** |
| Type 5 | 0.697 | 0.661 | 0.624 | 0.501 | 0.517 | **0.487** |

**Table 8.** The IT call center, call type 1: Coverage percentage of a 90% confidence predictor and of the two 5% tails for the test data set, for different queue sizes.

| Queue size | <5% | | | [5%, 95%] | | | >95% | | | Total obs |
|---|---|---|---|---|---|---|---|---|---|---|
| | QRF | LLOG | KDE | QRF | LLOG | KDE | QRF | LLOG | KDE | |
| 0 | 4.79 | 10.34 | 8.15 | 93.93 | 85.36 | 86.28 | 1.27 | 4.29 | 5.55 | 3778 |
| 1 | 5.93 | 9.60 | 7.60 | 90.54 | 86.81 | 87.13 | 3.52 | 3.57 | 5.25 | 3693 |
| 2 | 5.48 | 9.29 | 7.07 | 90.37 | 87.86 | 88.17 | 4.13 | 2.83 | 4.75 | 3407 |
| 3 | 6.22 | 9.38 | 7.32 | 89.92 | 87.57 | 87.78 | 3.85 | 3.04 | 4.88 | 2907 |
| 4 | 6.34 | 8.55 | 6.26 | 89.45 | 88.59 | 88.73 | 4.20 | 2.85 | 5.00 | 2380 |
| 5 | 6.94 | 8.44 | 6.25 | 89.14 | 88.17 | 88.95 | 3.91 | 3.37 | 4.78 | 2046 |
| ≥6 | 8.36 | 6.54 | 5.32 | 86.54 | 90.77 | 89.69 | 5.11 | 2.69 | 4.99 | 10837 |

**Table 9.** The IT call center, call type 1: Coverage percentage of a 80% PI and for the two 10% tails, for the test data set, for different queue sizes.

| Queue size | <10% | | | [10%, 90%] | | | >90% | | | Total obs |
|---|---|---|---|---|---|---|---|---|---|---|
| | QRF | LLOG | KDE | QRF | LLOG | KDE | QRF | LLOG | KDE | |
| 0 | 10.08 | 14.74 | 12.73 | 84.20 | 78.22 | 76.83 | 5.72 | 7.04 | 10.45 | 3778 |
| 1 | 11.40 | 14.32 | 12.48 | 79.53 | 78.45 | 77.23 | 9.07 | 7.23 | 10.29 | 3693 |
| 2 | 11.24 | 14.31 | 12.15 | 79.63 | 77.92 | 77.90 | 9.13 | 7.77 | 9.95 | 3407 |
| 3 | 11.35 | 14.62 | 12.32 | 80.29 | 77.85 | 77.74 | 8.36 | 7.53 | 9.94 | 2907 |
| 4 | 11.64 | 13.53 | 11.39 | 79.87 | 79.12 | 78.61 | 8.49 | 7.35 | 10.00 | 2380 |
| 5 | 12.70 | 13.10 | 11.32 | 79.46 | 79.52 | 78.59 | 7.84 | 7.38 | 10.09 | 2046 |
| ≥6 | 13.89 | 13.37 | 10.36 | 77.14 | 78.54 | 80.19 | 8.97 | 8.09 | 9.45 | 10774 |

Tables 10 and 11 show a different story for call type 2. QRF clearly dominates in the left tail, whereas in the right tail, LLOG wins for small queue sizes and QRF catches up for larger queue sizes. Figure 7 compares the density functions given by LLOG and KDE from the training data set. There is significant fitting error when the queue size is short (0 to 2), but it improves when the queue is larger. Overall, the coverage accuracies from LLOG and KDE are relatively similar.

**Table 10.** The IT call center, call type 2: Coverage percentage of a 90% PI and of the 5% tail on each side for the test data set, for different queue sizes.

| Queue size | <5% | | | [5%, 95%] | | | >95% | | | Total obs |
|---|---|---|---|---|---|---|---|---|---|---|
| | QRF | LLOG | KDE | QRF | LLOG | KDE | QRF | LLOG | KDE | |
| 0 | 5.37 | 9.39 | 11.95 | 93.21 | 85.30 | 85.19 | 1.41 | 5.30 | 2.85 | 4450 |
| 1 | 6.45 | 11.46 | 13.77 | 90.65 | 83.82 | 83.30 | 2.88 | 4.71 | 2.91 | 3672 |
| 2 | 7.05 | 13.19 | 16.53 | 89.53 | 83.01 | 81.03 | 3.41 | 3.79 | 2.42 | 2637 |
| 3 | 8.28 | 15.96 | 18.78 | 88.01 | 80.00 | 78.50 | 3.70 | 4.03 | 2.70 | 1810 |
| 4 | 7.11 | 12.76 | 13.53 | 88.07 | 81.95 | 82.26 | 4.81 | 5.27 | 4.20 | 1308 |
| 5 | 6.92 | 14.18 | 14.18 | 87.48 | 81.67 | 82.12 | 5.58 | 4.13 | 3.68 | 895 |
| ≥6 | 7.65 | 13.21 | 11.67 | 87.61 | 84.29 | 85.72 | 4.74 | 2.50 | 2.60 | 1961 |

Figure 8 shows Q-Q plot of the empirical distributions from the training data set and the test data set, for queue size from 2 to 5. These distribution of the training set and test set are different in the tails, especially when the error is negative (i.e., the delay is higher than predicted). The significant difference for $q_k = 3$ can be seen from the coverage values in Tables 10 and 11.

**Table 11.** The IT call center, call type 2: Coverage percentage of a 80% PI and of the 10% tail on each side for the test data set, for different queue sizes.

| Queue size | <10% | | | [10%, 90%] | | | >90% | | | Total obs |
|---|---|---|---|---|---|---|---|---|---|---|
| | QRF | LLOG | KDE | QRF | LLOG | KDE | QRF | LLOG | KDE | |
| 0 | 11.98 | 12.90 | 13.65 | 82.22 | 77.51 | 67.59 | 5.80 | 9.60 | 18.76 | 4450 |
| 1 | 12.23 | 16.04 | 19.20 | 80.28 | 75.52 | 74.65 | 7.49 | 8.44 | 6.15 | 3672 |
| 2 | 14,26 | 18.54 | 21.05 | 78.27 | 74.59 | 74.52 | 7.47 | 6.86 | 4.44 | 2637 |
| 3 | 16.08 | 21.22 | 23.37 | 76.08 | 71.22 | 71.16 | 7.85 | 7.57 | 5.47 | 1810 |
| 4 | 13.23 | 17.43 | 17.97 | 76.76 | 73.39 | 74.77 | 10.02 | 9.17 | 7.26 | 1308 |
| 5 | 12.74 | 19.89 | 19.89 | 78.10 | 71.51 | 73.30 | 9.16 | 8.60 | 6.82 | 895 |
| ≥6 | 14.13 | 21.62 | 20.49 | 77.72 | 72.97 | 74.29 | 8.28 | 5.41 | 5.28 | 1961 |

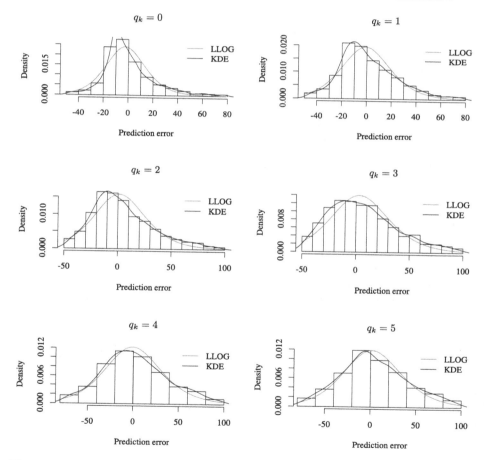

**Fig. 7.** The IT call center, call type 2: Density function of the prediction errors for LLOG and KDE, conditional on the queue size.

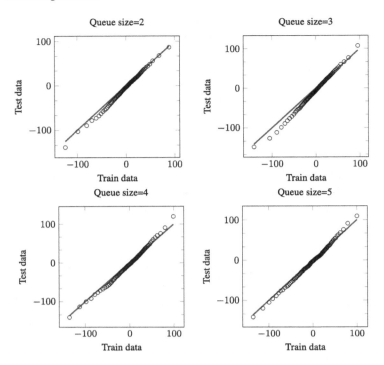

**Fig. 8.** The IT call center, call type 2: Q-Q plot of the empirical distributions between the training data set and the test data set.

## 7   Conclusion

We discussed and compared empirically different predictors for the waiting time of a call (or customer) when this call arrives at a multi-skill call center. The more accurate predictors among those examined are the predictors based on regression methods and automatic learning, and more specifically the predictors defined by deep multilayer neural networks, at least when enough data is available. We also examined different ways of modeling the distribution of the prediction error and estimating its density and its quantiles. We tried fitting known parametric distributions as well as a KDE to the observations of this prediction error conditional on the queue length, and the KDE usually gives a better fit even for the independent test data. But another non-parametric method named QRF, based on the random forest methodology, gave the best results in the majority of cases and performed reasonably well in general. On the flip side, this method is computationally more intensive. Suggestions for follow-up work include studying large call centers in which the waiting times are much longer, studying wait time predictions in other types of service systems (e.g., healthcare clinics), and trying to find better methods to estimate the density of the wait times.

**Acknowledgment.** This work was supported by grants from NSERC-Canada, Hydro-Québec, and a Canada Research Chair to P. L'Ecuyer. Ger Koole (VU Amsterdam) helped by providing data.

# References

1. Ang, E., Kwasnick, S., Bayati, M., Plambeck, E., Aratow, M.: Accurate emergency department wait time prediction. Manuf. Serv. Oper. Manag. **18**(1), 141–156 (2016)
2. Armony, M., Shimkin, N., Whitt, W.: The impact of delay announcements in many-server queues with abandonments. Oper. Res. **57**, 66–81 (2009)
3. Bengio, Y.: Practical recommendations for gradient-based training of deep architectures. Neural Netw. Tricks Trade **7700**, 437–478 (2012)
4. Bengio, Y., Courville, A.C., Vincent, P.: Unsupervised feature learning and deep learning: a review and new perspectives (2012). http://arxiv.org/abs/1206.5538
5. Bergstra, J., Bengio, Y.: Random search for hyper-parameter optimization. J. Mach. Learn. Res. **13**, 281–305 (2012)
6. Breiman, L.: Random forests. Mach. Learn. **45**(1), 5–32 (2001)
7. Cooper, R.B.: Introduction to Queueing Theory, 2nd edn. North-Holland, New York (1981)
8. de Boor, C.: A Practical Guide to Splines. Applied Mathematical Sciences Series, vol. 27. Springer, New York (1978)
9. Degenhardt, F., Seifert, S., Szymczak, S.: Evaluation of variable selection methods for random forests and omics data sets. Brief. Bioinform. **20**(2), 492–503 (2017)
10. Ding, R., McCarthy, M.L., Desmond, J.S., Lee, J.S., Aronsky, D., Zeger, S.L.: Characterizing waiting room time, treatment time, and boarding time in the emergency department using quantile regression. Acad. Emerg. Med. **17**(8), 813–823 (2010)
11. Dong, J., Yom Tov, E., Yom Tov, G.: The impact of delay announcements on hospital network coordination and waiting times. Manag. Sci. **65**(5), 1949–2443 (2018)
12. Friedman, J., Hastie, T., Tibshirani, R., Narasimhan, B., Simon, N., Qian, J.: R Package glmnet: Lasso and Elastic-Net Regularized Generalized Linear Models (2019). https://CRAN.R-project.org/package=glmnet
13. Friedman, J., Hastie, T., Tibshirani, R.: Regularization paths for generalized linear models via coordinate descent. J. Stat. Softw. **33**(1), 1–22 (2010)
14. Glorot, X., Bordes, A., Bengio, Y.: Deep sparse rectifier neural networks. In: Gordon, G., Dunson, D., Miroslav (eds.) Proceedings of the Fourteenth International Conference on Artificial Intelligence and Statistics. Proceedings of Machine Learning Research, vol. 15, pp. 315–323 (2011)
15. Goodfellow, I., Bengio, Y., Courville, A.: Deep Learning. MIT Press (2016). http://www.deeplearningbook.org
16. Goodfellow, I., et al.: Pylearn2: a machine learning research library, August 2013
17. Gross, D., Harris, C.M.: Fundamentals of Queueing Theory, 3rd edn. Wiley, New York (1998)
18. Gulcehre, C., Bengio, Y.: Knowledge matters: importance of prior information for optimization. J. Mach. Learn. Res. **17**, 1–32 (2016)
19. Hörmann, W., Leydold, J., Derflinger, G.: Automatic Nonuniform Random Variate Generation. Springer, Berlin (2004)
20. Ibrahim, R., L'Ecuyer, P., Shen, H., Thiongane, M.: Inter-dependent, heterogeneous, and time-varying service-time distributions in call centers. Eur. J. Oper. Res. **250**, 480–492 (2016)
21. Ibrahim, R., Whitt, W.: Real-time delay estimation based on delay history. Manuf. Serv. Oper. Manag. **11**, 397–415 (2009)

22. Ibrahim, R., Whitt, W.: Real-time delay estimation in overloaded multiserver queues with abandonments. Manag. Sci. **55**(10), 1729–1742 (2009)
23. Ibrahim, R., Whitt, W.: Delay predictors for customer service systems with time-varying parameters. In: Proceedings of the 2010 Winter Simulation Conference, pp. 2375–2386. IEEE Press (2010)
24. Ibrahim, R., Whitt, W.: Real-time delay estimation based on delay history in many-server service systems with time-varying arrivals. Prod. Oper. Manag. **20**(5), 654–667 (2011)
25. James, G., Witten, D., Hastie, T., Tibshirani, R.: An Introduction to Statistical Learning, with Applications in R. Springer, New York (2013)
26. Kleinrock, L.: Queueing Systems, vol. 1. Wiley, New York (1975)
27. Kursa, M.B., Rudnicki, W.R.: Feature selection with the Boruta package. J. Stat. Softw. **36**, 1–13 (2010)
28. LeCun, Y., Bengio, Y., Hinton, G.: Deep learning. Nature **521**, 436–444 (2015)
29. Meinshausen, N.: Quantile regression forests. J. Mach. Learn. Res. **7**, 983–999 (2006)
30. Nakibly, E.: Predicting waiting times in telephone service systems. Master's thesis, Technion, Haifa, Israel (2002)
31. Scott, D.W.: Multivariate Density Estimation, 2nd edn. Wiley, Hoboken (2015)
32. Senderovich, A., Weidlich, M., Gal, A., Mandelbaum, A.: Queue mining for delay prediction in multi-class service processes. Inf. Syst. **53**, 278–295 (2015)
33. Silverman, B.: Density Estimation for Statistics and Data Analysis. Chapman and Hall, London (1986)
34. Sun, Y., Teow, K.L., Heng, B.H., Ooi, C.K., Tay, S.Y.: Real-time prediction of waiting time in the emergency department, using quantile regression. Ann. Emerg. Med. **60**(3), 299–308 (2012)
35. Thiongane, M., Chan, W., L'Ecuyer, P.: Waiting time predictors for multiskill call centers. In: Proceedings of the 2015 Winter Simulation Conference, pp. 3073–3084. IEEE Press (2015)
36. Thiongane, M., Chan, W., L'Ecuyer, P.: New history-based delay predictors for service systems. In: Proceedings of the 2016 Winter Simulation Conference, pp. 425–436. IEEE Press (2016)
37. Thiongane, M., Chan, W., L'Ecuyer, P.: Delay predictors in multi-skill call centers: an empirical comparison with real data. In: Proceedings of the International Conference on Operations Research and Enterprise Systems (ICORES), pp. 100–108. SciTePress (2020). https://www.scitepress.org/PublicationsDetail.aspx?ID=QknUuVhZF/c=&t=1
38. Tibshirani, R.: Regression shrinkage and selection via the LASSO. J. R. Stat. Soc. Ser. B (Methodol.) 267–288 (1996)
39. Whitt, W.: Predicting queueing delays. Manag. Sci. **45**(6), 870–888 (1999)
40. Wood, S.N.: Generalized Additive Models: An Introduction with R, 2nd edn. Chapman and Hall/CRC Press, Boca Raton (2017)
41. Wood, S.N.: R Package MGCV: Mixed GAM Computation Vehicle with Automatic Smoothness Estimation (2019). https://CRAN.R-project.org/package=mgcv

# Reporting and Comparing Rates of Personnel Flow

Etienne Vincent[1]([✉]) [iD], Stephen Okazawa[1] [iD], and Dragos Calitoiu[2] [iD]

[1] Centre for Operational Research and Analysis, Department of National Defence, 101 Colonel By Drive, Ottawa, ON K1A 0K2, Canada
`{Etienne.Vincent,Stephen.Okazawa}@Forces.gc.ca`
[2] Director General Military Personnel Research and Analysis, Department of National Defence, 101 Colonel By Drive, Ottawa, ON K1A 0K2, Canada
`Dragos.Calitoiu@Forces.gc.ca`

**Abstract.** In Personnel Operations Research (OR), rates of personnel flow, such as rates of attrition, promotion and transfer, are commonly reported, compared and modelled. However, different practitioners often use different formulas to do this, and these formulas have often been defined without justification. This chapter rigorously defines a novel formula for measuring personnel flow rates in order to solidify the foundations of Personnel OR and encourage consistency and comparability in reported metrics. We refer to this new definition as the *general formula* for rates of personnel flow. The general formula is justified by its properties, but also by analogy with the field of investment performance measurement, where a similar formula is known as the Time Weighted Rate of Return. We also derive the related internal rate of personnel flow, for applications in forecasting. Finally, we develop various approximations to the general formula, the accuracy of which we then empirically evaluate using Canadian Armed Forces Human Resources data.

**Keywords:** Personnel operations research · Attrition rate · Promotion rate · Investment performance measurement

## 1 Introduction

This chapter is an expanded version of our paper published in the proceedings of the 10th International Conference on Operations Research and Enterprise Systems [1]. We add to that paper a clear illustration of the fact that the choice of formula used to measure personnel flow rates matters (Sect. 2), an additional approximation formula (Eq. (16)) that we will empirically compare to the ones that we had previously introduced, and a real-world example that highlights a common pitfall of relying exclusively on using rates to compare workforce sub-populations (Sect. 6).

Personnel Operations Research (OR) informs operational decisions through workforce modelling, and analytics derived from data found in Human Resources (HR) systems. Personnel flows are prevalent in Personnel OR. These include attrition, promotions and transfers. Rates of flows are commonly reported, used to compare various workforces or their sub-populations, used to inform forecasts, and used as inputs within workforce

G. H. Parlier et al. (Eds.): ICORES 2020/2021, CCIS 1623, pp. 107–127, 2022.
https://doi.org/10.1007/978-3-031-10725-2_6

models. However, different practitioners of Personnel OR often use different formulas to measure these rates. As pointed out by Noble, all agree that attrition rates are important and that their definition is self-evident, but then go on to give different definitions [2].

In this chapter, we propose to solidify the foundations of Personnel OR by introducing a formal definition for personnel flow rates, which we call the *general formula*. As such, we generalize and improve on work by Okazawa [3], the only previous attempt of which we are aware to theoretically justify a formula. We will also present practical approximation formulas for our general formula, and will empirically compare their accuracy.

## 2   Why the Choice of Rate Formula Matters

The treatment of personnel flow rates, a concept that is fundamental in Personnel OR, is often imprecisely treated by authors. For example, Bartholomew, Forbes and McClean [4], a foundational reference to Personnel OR, provides two alternative definitions for attrition rates, but does not attempt to provide theoretical justification – they are offered as self-evident. A commonly held view in the Personnel OR community is that many alternative rate definitions are equally valid, as long as they are applied consistently. This section aims to dispel this view by providing a simple example to show that different commonly seen definitions can lead to diametrically opposed conclusions.

Attrition rates are most-often defined by taking the flow volume, and dividing it by the size of the population. Differences then emerge in how to measure the population, since it usually varies over the period in question. Two choices (among others) are to take the initial population plus half of the external flows to and from the population (called the *half-intake* rate, which was introduced in [3]), or the average population over the period in question (called *central wastage* rate in [4]). We will now apply these definitions to the synthetic examples shown in Fig. 1.

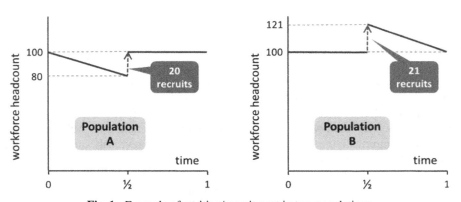

**Fig. 1.** Example of attrition/recruitment in two populations.

Figure 1 tracks two populations (A and B) over a period of unit duration. Both populations start with 100 employees. Over the first half, 20 employees gradually depart population A, but are instantaneously replaced by incoming recruits at the half point. Over the second half, that population remains unchanged. Population B is instead unchanged in the first half, while 21 recruits are gained at the half-way point and the same number of employees is gradually lost over the second half. These examples are artificial, but not completely implausible. When applied to populations A and B, the half-intake and central wastage rates shown in Table 1 are obtained.

**Table 1.** Attrition rates calculated using two different formulas for the populations shown in Fig. 1.

| Attrition rate formula | Population A | Population B |
|---|---|---|
| Half-intake | 18.2% | 19.0% |
| Central wastage | 21.1% | 20.0% |

Notice that in Table 1 the half-intake formula shows Population B as experiencing higher attrition than A, while the central wastage formula leads to the opposite conclusion. This example is based on two specific formulas, but similar examples can be designed to show contrary conclusions from other pairs of formulas. It thus becomes clear that formally justifying our choice of formula will be necessary if important decisions are to rely on personnel flow rates.

# 3  Proportional Rates in Other Domains

Attrition, promotion and transfer rates are proportional rates. They measure flows as proportions of a population. For example, the attrition rate is the proportion that leaves the population in a given time period, while the promotion rate tracks the proportion of employees who are promoted. Similar proportional rates are found in other disciplines, and can be a source of inspiration to Personnel OR.

## 3.1  Demographics

Demographics makes extensive use of proportional rates. For example, divorce, emigration or mortality rates are similar to Personnel OR's attrition rates. However, demographic data often come from multiple disparate sources, unlike personnel data, which is generally from a single HR system. For example, divorces are promulgated by courts and tracked by justice systems, whereas counts of married couples come from censuses. This makes it impossible to track day-to-day changes in the married population, which would additionally require reconciliation of immigration and mortality data from yet other sources. Divorce rates, therefore, end up being presented as simple ratios between the numbers of divorces and census population.

Demographics nonetheless makes extensive use of standardized rates [5], which offer a solution to the problem of reconciling rates for sub-populations with the whole: a problem that is often seen in Personnel OR. For example, [6] presents a population where the men have a higher attrition rate than the women, but that when broken down by age, the women have a higher attrition rate in each of the age groups. This apparent contradiction is explained by the fact that the women in question were younger as a whole, and attrition was higher among younger employees (both women and men). Standardized rates offer a way of comparing these women and men normalized to a standard age distribution, an approach that could often be helpful but is seldom seen in Personnel OR. Instead, Personnel OR typically proceeds by segmenting workforces into their heterogeneous components [4, 7].

## 3.2  Subscription Services

Proportional rates are also used to report the subscriber churn in subscription services. For example, churn rates are typically an important measure of performance for wireless service providers, and are presented in their quarterly financial reports. As an example, AT&T reports its churn rates as the average over months of the number of subscribers who cancel their service each month divided by the number of subscribers at the beginning of the respective months [8]. There is, however, no industry standard definition, and different carriers may use different definitions for their churn rates. Furthermore, AT&T's measure might work in practice, but is not rigorous, since new subscribers are acquired during the month, and some cancelling subscribers might not have been there at the beginning of the month. Different intra-month patterns in subscriber flows could thus skew results and lead to false conclusions on the carrier's operational performance. Overall, rates reporting for subscription services is even less mature than in Personnel OR.

## 3.3  Internal Rate of Return

To our knowledge, Finance is the discipline where proportional rates are most mature. Interest rates and rates of return are foundational to that discipline. In particular, investment performance measurement shares important similarities with the reporting of rates in Personnel OR. Investment performance measurement is highly standardized by regulatory bodies that aim to achieve a fair comparison of the returns achieved by different investment firms. The remainder of this section concerns the treatment of rates in investment performance measurement, while we will subsequently develop analogous methods for use in the context of Personnel OR.

Figure 2 provides an illustration, reproduced from [1], which tracks the value of an investment account over a year. Initially, the account contains investments valued at $40K, which increase in value to $50K within three months, representing a 25% increase. At that point, $200K is transferred into the account. Over the next three months, the value of the account drops to $150K – a 40% reduction, before $100K is transferred out of the account. In the last six months of the year, the value of the account grows by 60% from $50K to $80K.

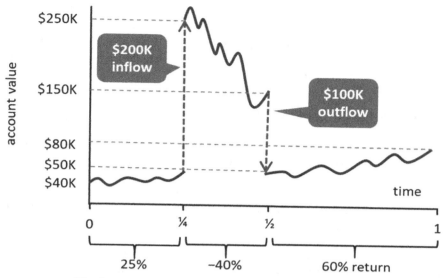

**Fig. 2.** Example tracking an investment account over a year.

The *internal rate of return* (IRR) is widely used in Finance. It is the effective rate that achieves the account's final value, when applied to the initial value and all deposits and withdrawals. For this example, the IRR is $-42.0\%$, as obtained by solving

$$80 = 40(1+r) + 200(1+r)^{\frac{3}{4}} - 100(1+r)^{\frac{1}{2}} \tag{1}$$

The Canadian Securities Administrators (CSA), along with many of its international counterparts, requires that the investment performance achieved within client accounts be reported using the IRR or related metrics [9]. This amounts to reporting the fixed interest rate that would have resulted in the same gain (or loss) by year-end as was achieved in aggregate by the particular securities held in the given account. This measure is intuitive, and informative for the account holders.

However, the IRR is not an appropriate performance measure for investment fund managers. To see this, consider that the IRR varies not only with the outcome of the fund manager's investment decisions, but also with the amounts deposited and withdrawn, which are not subject to the fund managers' control. For the example in Fig. 2, the poor timing of the external transfers coinciding with a sub-period of negative returns is largely responsible for the overall loss.

### 3.4  Time-Weighted Rate of Return

To measure the performance of fund managers, the *time-weighted rate of return* (TWRR) offers a measure that is invariant with respect to deposits and withdrawals. It is obtained by compounding rates of return over the sub-periods between each transfer. In our example, the TWRR is of 20%, obtained as

$$\frac{50}{40} \cdot \frac{150}{250} \cdot \frac{80}{50} - 1 \tag{2}$$

The TWRR amounts to the rate of return that would have resulted from a set invest-ment in the fund, subject only to changes in the value of the assets held in that fund (as if there had been no deposits or withdrawals). Use of the TWRR is mandated by the Char-tered Financial Analyst (CFA) Institute's Global Investment Performance Standards for reporting investment fund returns [10, 11].

## 4   Defining Personnel Flow Rates

If the investment flows shown in Fig. 2 were reinterpreted to depict personnel flows, they would be deemed unrealistic. A workforce is unlikely to be subject to such dra-matic outflows and inflows, and this somewhat explains why the definition of personnel flow rates has not been given as much urgency as its financial counterpart. Neverthe-less, for a typical employer, recruitment volumes vary from year to year. Furthermore, organizations are occasionally affected by atypical events, such as mergers, spin-offs, or downsizing.

When personnel flow rates are being measured and reported, it is often in order to compare them to either a standard (e.g. is the level of attrition high or low?), or to other rates (e.g. is a sub-population being promoted more or less than others?). The basic reason that rates are used is to have a measure that is invariant to the size of the population in question. Similarly, invariance to uncontrolled external flows is often desirable. Invariance to uncontrolled factors isolates the phenomenon of interest, such that the comparison is as fair as possible. An approach similar to the TWRR from invest-ment performance measurement thus appears relevant. The performance of leaders with respect to their management of a workforce will be best assessed through a measure-ment that is independent of factors outside their control, such as prescribed cutbacks, or mergers with other units. For Personnel OR, we refer to the TWRR as the *general formula* for personnel flow rates. We first defined this formula specifically for attrition rates, in a Canadian Department of National Defence internal report [12].

The general formula can be used to measure rates of attrition, promotions, transfers, and all of their variations. Among these, attrition is the most commonly reported, and here will be understood to include departures for any reason (resignation, retirement, dismissal, etc.). To simplify the remainder of this chapter, we will often describe concepts in terms of attrition rates, but we remind the reader that the discussion usually also applies to other proportional rates of personnel flow.

### 4.1   The General Formula

In order to illustrate the general formula with an example, Fig. 3 (reproduced from [1]) depicts an imaginary workforce's headcount over a year. For illustrative purposes, attrition is atypically high in this example.

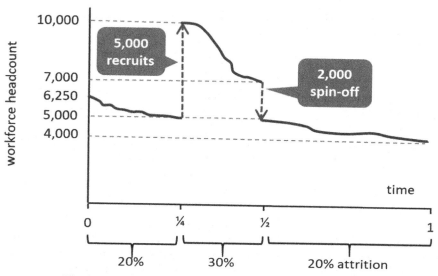

**Fig. 3.** Example tracking headcount in an imaginary workforce.

The headcount starts at 6,250 and gradually decreases. After the third month, 5,000 recruits show up. Then, at the six month mark, 2,000 employees are transferred out as the result of a spin-off (an outflow that is not to be counted as attrition). In the end, 4,000 employees are left. In this example, the general formula rate of attrition is 55.2%, as obtained by compounding retention rates from the three sub-periods that are free of other flows:

$$1 - \frac{5,000}{6,250} \cdot \frac{7,000}{10,000} \cdot \frac{4,000}{5,000} \tag{3}$$

As with the TWRR, the general formula rate does not vary with the timing of non-attrition flows – only the relative changes occurring between those non-attrition flows matter. Typically, HR data is captured at a daily resolution, with inflows and outflows occurring between work days. The general formula can then be concisely expressed as a compounding of daily rates:

$$1 - \alpha = \prod_{i=1}^{n} \frac{p(i)}{p(i) + a(i)} \tag{4}$$

where $\alpha$ is the rate being measured, $n$ the number of days in the period of interest, $p(i)$ the headcount at the end of the $i^{\text{th}}$ day, and $a(i)$ the magnitude of the relevant personnel flow on that day (e.g. attrition).

We will now highlight two properties of the general formula. The first is that when $a(i)$ is the only flow impacting the workforce, $p(i) + a(i) = p(i-1)$ for all $i$, which leads to:

$$1 - \alpha = \frac{p(n)}{p(0)} \tag{5}$$

If a is the sum of all a($i$) over the period, we also get $p(n) = p(0) - a$, such that Eq. (5) becomes

$$\alpha = \frac{a}{p(0)} \tag{6}$$

This happens to be one of the common definitions of the attrition rate (*called transition wastage* in [4]), but which we now reinterpret as a special case of the general formula applicable when attrition is the only flow.

A second property to highlight is that given sub-period rates $\alpha_i$ (i.e. daily, or monthly rates), the general formula rate can be obtained by compounding:

$$1 - \alpha = \prod_i (1 - \alpha_i) \tag{7}$$

This multiplicative property directly follows from Eq. (4).

In [12], instead of using the TWRR as a starting point, we had undertaken to derive a general formula for attrition from desirable properties. We had settled on two desirable properties: multiplicativity (i.e. Eq. (7)) and simple division by the initial population in the absence of other flows (i.e. Eq. (6)). The reader may note that the general formula (Eq. (4)) can be directly derived from these properties by taking $\alpha_i$ as daily rates and computing these from Eq. (6) under the assumption that all flows occur between workdays. It is only later that we noticed parallels with investment performance measurement, offering further justification for preferring this attrition rate definition.

### 4.2   Internal Rate of Personnel Flow

In investment performance measurement, besides the TWRR, we saw that the IRR is also of importance. Similarly, although we prefer the general formula when it comes to reporting and comparing flows, a counterpart of the IRR can have important applications in forecasting and modelling. We will now show how it may be derived from the general formula under additional constraints.

Figure 4 tracks a workforce undergoing a single non-attrition flow of magnitude $x$, perhaps the arrival of a cohort of new hires, occurring at time $k$.

From Eq. (5), the attrition rate over the sub-periods without external flows are obtained from the ratios of start/end headcounts as

$$1 - \alpha_{[0,k]} = \frac{p(k) - x}{p(0)} \tag{8}$$

and

$$1 - \alpha_{[k,1]} = \frac{p(1)}{p(k)} \tag{9}$$

with $p(k)$ denoting the population immediately before the non-attrition flow of magnitude $x$. From Eq. (7), attrition over the entire period [0, 1] is then

$$1 - \alpha = \frac{p(k) - x}{p(0)} \cdot \frac{p(1)}{p(k)} \tag{10}$$

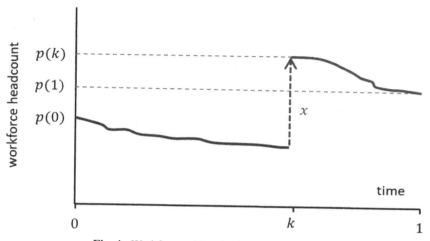

**Fig. 4.** Workforce with a single non-attrition flow.

Substituting Eq. (10) into the following easily verifiable identity

$$p(1) = p(0) \cdot \left[ \frac{p(k) - x}{p(0)} \cdot \frac{p(1)}{p(k)} \right] + x \cdot \frac{p(1)}{p(k)} \tag{11}$$

we obtain

$$p(1) = p(0) \cdot (1 - \alpha) + x \cdot (1 - \alpha_{[k,1]}) \tag{12}$$

Note that Eq. (12) separates the effect of attrition on $p(0)$ from its effect on $x$, and may be generalized to any number of non-attrition flows. This suggests a naïve approach to forecasting. Given $p(0)$, and a planned future non-attrition flow $x$ at time $k$ (e.g. the planned intake of recruits), Eq. (12) will give us a forecasted value for $p(1)$, but only if we also have foreknowledge of $\alpha_{[k,1]}$.

As the value of $\alpha_{[k,1]}$ cannot be know at time 0, a reasonable unbiased assumption might be that attrition will happen at the same pace over the interval $[k, 1]$, as it will over the full $[0, 1]$. This amounts to setting

$$1 - \alpha_{[k,1]} \cong (1 - \alpha)^{1-k} \tag{13}$$

Then, Eq. (12) becomes

$$p(1) \cong p(0) \cdot (1 - \alpha) + x \cdot (1 - \alpha)^{1-k} \tag{14}$$

which, for an arbitrary number of external flows $x_i$ occurring at times $k_i$, generalizes as

$$p(1) \cong p(0) \cdot (1 - \alpha) + \sum_{i=1}^{n} x_i \cdot (1 - \alpha)^{1-k_i} \tag{15}$$

Now, with an expected rate of attrition $\alpha$, we can forecast future workforce head-counts. We call Eq. (15) the *internal rate of personnel flow*. It may be understood as a model derived from the general formula under the assumption that attrition will progress at a fixed pace. This assumption is unlikely to be strictly true in practice, but is reasonable when nothing is known about the actual attrition pattern, such as when forward-projecting a rate in order to predict future headcounts. The internal rate of personnel flow is thus a special case of the general formula.

In investment performance measurement, the TWRR and IRR can give very different outputs, as was the case for the example in Fig. 2. In Personnel OR, on the other hand, the rates from the general formula and internal rate of personnel flow are typically much closer. This is because personnel flows tend to occur at a steadier pace, are never negative, and tend to be small relative to the headcount.

## 5    Rate Approximations

The general formula (Eq. (4)) is our preferred metric for measuring, reporting and comparing attrition, promotion, and transfer flows. However, it is often inconvenient or even impossible to apply directly in practice. Let us illustrate with an example.

Consider wanting to measure the attrition rate among Royal Canadian Air Force pilots. First, it should be easy to extract the necessary daily attrition counts from the relevant HR System logs. Then, the general formula also requires daily headcounts for the pilot population. Daily headcounts tend to be harder to obtain because they must be computed from the logs that track all personnel movements in and out of the pilot occupation, such as recruitment, attrition and transfers. It is possible to coordinate all of these transactions, but not straightforward when some may occur simultaneously, and when the logs contain inconsistencies. Nevertheless, code can be developed to do it. Now, consider that we may want to break down the pilot population into segments to compare attrition between ranks, wings, and, age groups. Obtaining daily headcounts then involves tracking promotions, postings, and birthdays. This is likely still possible, but quickly becomes unwieldy. Furthermore, it is possible that the data needed to define some segments of interest is simply not tracked at a daily resolution.

If it was possible to approximate the general formula in a way that did not require daily headcounts, the task of measuring and comparing rates would be greatly simplified. In this section, we will derive four such approximation formulas, and then compare them by applying them to real data.

We are looking for formulas for estimating $\alpha$ from $p(0)$ (initial headcount), $p(1)$ (headcount at the end of the period – usually a year) and a (total attrition volume over the period). When applying such approximations, the attrition volume (a) will be easy to extract from attrition transaction logs, while $p(0)$ and $p(1)$ can be obtained from precomputed annual workforce snapshots that list all employees with their relevant attributes (e.g. rank, occupation, work location, age, sex, etc.) at the end of each year.

### 5.1    Middle-Intake Approximation

Given $p(0)$, $p(1)$ and a, the value taken by the general formula ($\alpha$) varies with the path taken by the headcount as it progresses from $p(0)$ to $p(1)$. In other words, by postulating a distribution of attrition events relative to the non-attrition flows, we obtain a potential value for $\alpha$. Our first three approximation formulas are obtained by directly selecting such a distribution of attrition relative to other flows.

For our first attempt to approximate $\alpha$, we lump all non-attrition flows together into a single event occurring half-way along the time interval, with half of a placed before that event and the other half after. This is illustrated in Fig. 5.

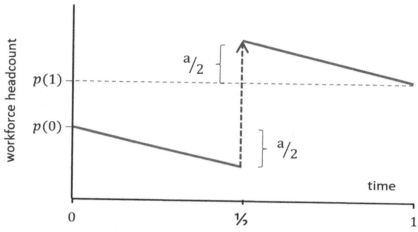

**Fig. 5.** Flows resulting in the middle-intake approximation.

The general formula, applied to this assumed distribution, yields the product of two ratios corresponding to the sub-periods before and after the interval mid-point. Over each of these sub-periods, there are no non-attrition flows, so measuring attrition over them is trivial. Over the first sub-period, attrition brings the population from $p(0)$ to $p(0) - a/2$, while it goes from $p(1) + a/2$ to $p(1)$ over the second period.

$$1 - \alpha \cong 1 - \alpha_{MI} = \frac{p(0) - \frac{a}{2}}{p(0)} \cdot \frac{p(1)}{p(1) + \frac{a}{2}} \tag{16}$$

We call $\alpha_{MI}$ the *middle-intake* approximation formula, as it is derived by assuming that non-attrition flows (which in many situations are largely made up of intake into the population) occur half-way through the attrition.

## 5.2  Half-Intake Approximation

To derive a second approximation formula, we first define

$$x = p(1) - p(0) + a \tag{17}$$

as the net non-attrition flow in or out of the workforce. Then, we assume that half of $x$ occurs before any attrition is observed, while the other half comes after all attrition has been recorded, as illustrated in Fig. 6.

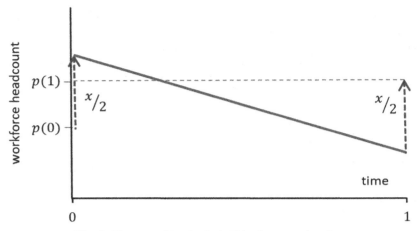

**Fig. 6.** Flows resulting in the half-intake approximation.

Under these assumptions, attrition takes the population from $p(0) + x/2$ down to $p(1) - x/2$. We thus get

$$1 - \alpha \cong 1 - \alpha_{HI} = \frac{p(1) - \frac{x}{2}}{p(0) + \frac{x}{2}} \tag{18}$$

With Eq. (17), this can be re-organized into the more commonly seen form

$$\alpha_{HI} = \frac{a}{p(0) + \frac{x}{2}} \tag{19}$$

In the investment performance measurement literature, Eq. (19) is known as the Simple Dietz method. In that context, it was originally derived to approximate the returns from uncompounded interest [13]. In our Personnel OR context, we call Eq. (19) the *half-intake* formula, as it is obtained by adding half of the net non-attrition flows (including intake) to the denominator.

Our original goal was to derive an approximation using only $p(0)$, $p(1)$ and a. To do this, we again apply Eq. (17) to obtain

$$\alpha_{HI} = \frac{2a}{p(0) + p(1) + a} \tag{20}$$

The half-intake formula was first derived, in the context of Personnel OR, by Okazawa [3]. However, that derivation did not use the general formula as a starting point. Rather, it was derived as an approximation of the uniform Taylor approximation (seen next subsection). Following that initial derivation, the half-intake formula has become the *de facto* standard for attrition rate measurement in the Canadian Armed Forces.

### 5.3 Uniform Taylor Approximation

As we will see, the middle-intake and half-intake formulas will prove to effectively approximate the general formula in practice. However, the distributions that led to these approximations, shown in Fig. 5 and Fig. 6, are highly artificial. A more realistic assumption on how a number of personnel flows are actually distributed in practice is that they are spread fairly evenly across time. To derive an approximation based on this assumption, we will assume uniformly distributed net non-attrition flows, together with a constant pace of attrition. As we saw previously, saying that attrition occurs at a constant pace amounts to having it conform to the internal rate of personnel flow formula (Eq. (15)). When we distribute $x$ evenly across time within the internal rate of personnel flow formula, we get

$$p(1) \cong p(0) \cdot (1 - \alpha) + \sum_{i=1}^{x} (1 - \alpha)^{1 - \frac{i}{x+1}} \tag{21}$$

In order to benefit from classical numerical approximations for continuous functions (decomposition into series), we map Eq. (21) to a continuous form:

$$p(1) \cong p(0) \cdot (1 - \alpha) + x \int_{0}^{1} (1 - \alpha)^{1-t} dt \tag{22}$$

which, through integration, becomes

$$p(1) \cong p(0) \cdot (1 - \alpha) + x \cdot \frac{-\alpha}{\ln(1 - \alpha)} \tag{23}$$

In order to avoid a numerical solution for $\alpha$, we make use of the Taylor series

$$\frac{-\alpha}{\ln(1 - \alpha)} \cong 1 - \frac{\alpha}{2} - \frac{\alpha^2}{12} - \frac{\alpha^3}{24} - \cdots \tag{24}$$

To obtain a direct approximation formula, we only keep the quadratic terms of Eq. (24), such that when we substitute into Eq. (23), the result is a quadratic polynomial:

$$p(1) \cong p(0) \cdot (1 - \alpha) + x \cdot \left( 1 - \frac{\alpha}{2} - \frac{\alpha^2}{12} \right) \tag{25}$$

which solves as

$$\alpha \cong \alpha_{UT} = \begin{cases} \frac{p(0)-p(1)}{p(0)}, & \text{when } x = 0 \\ \frac{p(0)+\frac{x}{2}-\sqrt{(p(0)+\frac{x}{2})^2+\frac{x}{3}(p(0)+x-p(1))}}{-x/6}, & \text{otherwise} \end{cases} \tag{26}$$

We refer to this formula as the *uniform Taylor* approximation.

Notice that if only the linear terms of Eq. (24) had been kept, an alternative derivation of the half-intake formula would have resulted, which is how this formula was derived in [3].

## 5.4 Mean Continuous Approximation

This last approximation formula will again be derived from the internal rate of personnel flow. Following an approach that is often employed in the field of Finance, we first convert the periodically compounding rate ($\alpha$) into a continuously compounding rate ($\gamma$), by defining

$$\gamma = -\ln(1 - \alpha) \tag{27}$$

When the conversion is applied to Eq. (15), the internal rate of personnel flow formula becomes

$$p(1) \cong p(0) \cdot e^{-\gamma} + \sum_{i=1}^{n} x_i \cdot e^{-\gamma(1-k_i)} \tag{28}$$

Then, by substituting Eq. (28) into Eq. (17), we get:

$$a = p(0) + x - p(1) \cong p(0) \cdot (1 - e^{-\gamma}) + \sum_{i=1}^{n} x_i \cdot \left(1 - e^{-\gamma(1-k_i)}\right) \tag{29}$$

On the other hand, from Eq. (28), we can also obtain the mean headcount over the period by separating the effect of attrition on the initial headcount $p(0)$ from its effect on each of the non-attrition flows $x_i$ as:

$$\bar{p} \cong \int_0^1 p(0) \cdot e^{-\gamma t} dt + \sum_{i=1}^{n} \int_{k_i}^1 x_i \cdot e^{-\gamma(1-t)} dt$$

$$= \frac{p(0) \cdot (1 - e^{-\gamma})}{\gamma} + \sum_{i=1}^{n} \frac{x_i \cdot \left(1 - e^{-\gamma(1-k_i)}\right)}{\gamma} \tag{30}$$

Now, notice that when we divide Eq. (29) by Eq. (30), we obtain:

$$\frac{a}{\bar{p}} \cong \gamma \tag{31}$$

We call Eq. (31) the *mean continuous* approximation formula because it represents a continuously compounding rate, and is obtained by dividing by the mean population. In Sect. 2, we noted that an attrition rate definition that is commonly found in the non-technical HR literature is the attrition volume divided by the mean population, so essentially Eq. (31). As we will show, this indeed leads to a useful approximation of the general formula for the attrition rate, but only when the resulting rate is interpreted as continuously compounding. Using Eq. (27), $\gamma$ may be converted back to an annually compounding rate to obtain

$$\alpha \cong 1 - e^{-a/\bar{p}} \tag{32}$$

Notice that no assumption was thus far made about the pattern of non-attrition flows. As such, Eq. (32) is really just a reformulation of the internal rate (Eq. (15)). This is noteworthy because Eq. (15) and the IRR used in Finance are generally thought of as requiring a numerical solution, whereas Eq. (32) does not. Nevertheless, Eq. (32) requires $\bar{p}$, which is not typically readily available.

Our goal is still to obtain an approximation relying only on $p(0)$, $p(1)$ and a. The simple estimate of $\bar{p}$ that may be obtained from these would then be $(p(0) + p(1))/2$, which gives us:

$$\alpha \cong \alpha_{MC} = 1 - \exp\left(\frac{-2a}{p(0) + p(1)}\right) \tag{33}$$

We have investigated other choices for $\bar{p}$, such as one that assumes an exponential trajectory between $p(0)$ and $p(1)$, but these other choices produced worse results in our empirical tests. Determining an appropriate $\bar{p}$ from $p(0)$ and $p(1)$ is not obvious, as the true $\bar{p}$ need not even be a value between $p(0)$ and $p(1)$. For example, if recruits all come in around the same time, but attrition is spread across the year, the headcount follows an annual cycle that peaks during the season when the recruits arrive. If that recruit arrival season is soon after year start, $p(0)$ and $p(1)$ will be near the annual low points, and thus below the annual average population. Awareness of the expected annual pattern for a specific workforce could lead to a better estimate of $\bar{p}$, but is beyond our current scope of looking for generic estimation formulas.

## 5.5  Empirical Evaluation Framework

Now that we have derived four candidate approximation formulas, we are ready to empirically test, on real data, how well they match the values obtained directly from the general formula. For this, we employ the same dataset as in [1]. That dataset includes 10 years of Canadian Armed Forces Regular Force HR data ending in 2019. We decided not to include more recent data, as these would substantially diverge from typical patterns due to the impact of the COVID-19 pandemic.

The dataset tracks 32 segments of the workforce (separated according to age (below/above 40), Occupational Authority (Army, Navy, Air Force and joint trades), and rank (junior/senior officers/non-commissioned members). On each segment, five personnel flows were tracked: attrition, medical releases, component transfers to the Primary Reserve Force, promotions, and the occupation transfers of members transferring from one Occupational Authority to another. This gives us 1,600 tests in total (10 years × 32 segments × 5 types of flows).

In each of the 1,600 tests, we measured annual rates obtained directly from the general formula on the full daily-resolution data. We also applied the four approximation formulas using only the annual values of a, $p(0)$ and $p(1)$. The approximations considered are the middle-intake ($\alpha_{MI}$ from Eq. (16)), half-intake ($\alpha_{HI}$ from Eq. (20)), uniform Taylor ($\alpha_{UT}$ from Eq. (26)) and mean continuous ($\alpha_{MC}$ from Eq. (33)).

## 5.6  Evaluation Results

How accurately each of our approximation formulas matched the results obtained from the general formula is shown in Table 2. The percentages shown are the mean absolute differences between the rates directly derived from the general formulas and those derived from approximations. The first five rows are each based on 320 tests corresponding to one of the five flow types on the 32 segments and 10 years of data, while the last row is the mean absolute difference over all 1,600 tests.

**Table 2.** Mean absolute difference between direct application of the general formula and approximations.

|  | $\alpha_{MI}$ | $\alpha_{HI}$ | $\alpha_{UT}$ | $\alpha_{MC}$ |
|---|---|---|---|---|
| Attrition | 0.069866% | 0.068931% | 0.069231% | 0.069488% |
| Medical release | 0.022291% | 0.022225% | 0.022386% | 0.022508% |
| Component transfer | 0.008466% | 0.008456% | 0.008464% | 0.008459% |
| Promotion | 0.038522% | 0.038723% | 0.039441% | 0.039422% |
| Occupation transfer | 0.028012% | 0.027445% | 0.027347% | 0.027390% |
| Overall | 0.032914% | 0.032670% | 0.032885% | 0.032969% |

The first thing to note is that all four formulas provide very close approximations – within 0.07% of the direct application of the general formula, on average. Personnel flow rates are rarely reported at resolutions below a tenth of a percent, so the observed accuracy is satisfactory. Overall, the half-intake approximation was the most accurate, but only by a small margin.

Whereas Table 2 present the formulas' average accuracy, Table 3 shows the worst accuracy observed over all the tests. With this metric, the mean continuous approximation proves the most accurate, but again, only by a small margin. The middle-intake approximation is however more clearly outperformed by the others. Over all of the tests, it is noteworthy that all of the approximations obtained from the four formulas remain within 0.53% of the exact rates. For most applications, this is sufficient accuracy, while achieved at a considerable saving in terms of the effort that would be required to parse the necessary data. In particular, the worst approximation obtained for each formula was in 2011 in one of the smallest segments considered, which had an average headcount of 201 persons that year. Thus, the worst absolute differences slightly above 0.5% shown in Table 3 amount to differences of about one person due to the approximation.

**Table 3.** Maximum absolute difference between direct application of the general formula and approximations.

|  | $\alpha_{MI}$ | $\alpha_{HI}$ | $\alpha_{UT}$ | $\alpha_{MC}$ |
|---|---|---|---|---|
| Attrition | 0.525037% | 0.504028% | 0.506517% | 0.502548% |
| Medical release | 0.161480% | 0.159302% | 0.162011% | 0.163096% |
| Component transfer | 0.105358% | 0.097457% | 0.097575% | 0.097451% |
| Promotion | 0.328743% | 0.330949% | 0.354450% | 0.355539% |
| Occupation transfer | 0.480166% | 0.471317% | 0.467894% | 0.469390% |
| Overall | 0.525037% | 0.504028% | 0.506517% | 0.502548% |

## 5.7   Approximations Based on Monthly Data

In the previous section, we saw how the approximation formulas have yielded personnel flow rate approximations within 0.53% of those obtained directly from by the general formula in all of our 1,600 tests. Nevertheless, if a closer approximation was required, it can still be obtained with less effort than from the general formula by considering shorter sub-annual periods, and compounding their rates into annual ones. In this section, we will consider the exploitation of monthly data. We assume the availability of monthly values for a and the month start/end headcounts to compute monthly rate approximations using the four formulas. To obtain annual approximations from twelve monthly approximations, we compound them using:

$$1 - \alpha = \prod_{i=Jan}^{Dec} 1 - \alpha_i \tag{34}$$

Table 4 presents the mean absolute differences between approximations obtained in this way and the rates obtained directly from the general formula. By comparing Table 4 with Table 2, we see that consideration of monthly data results in substantially better accuracy. In this test, the best-performing approximation was found to be the uniform Taylor, but as before, only by a small margin.

**Table 4.** Mean absolute difference between direct application of the general formula and approximations derived from monthly data.

|  | $\alpha_{MI}$ | $\alpha_{HI}$ | $\alpha_{UT}$ | $\alpha_{MC}$ |
|---|---|---|---|---|
| Attrition | 0.014415% | 0.014164% | 0.014139% | 0.014129% |
| Medical release | 0.004632% | 0.004560% | 0.004558% | 0.004555% |
| Component transfer | 0.002099% | 0.002071% | 0.002070% | 0.002070% |
| Promotion | 0.015586% | 0.015304% | 0.015274% | 0.015279% |
| Occupation transfer | 0.008927% | 0.008601% | 0.008576% | 0.008592% |
| Overall | 0.009166% | 0.008974% | 0.008957% | 0.008959% |

Similarly to Table 3, Table 5 presents the worst-case absolute differences, but this time based on approximations from monthly data. Again, we see an improvement over approximations obtained from annual data. Now, the best-performing uniform Taylor approximation gives rates that are always with 0.24% of the exact values.

In practice, this consideration of monthly data has become our preferred method for efficiently obtaining sufficiently accurate personnel flow rates for the Canadian Armed Forces. We maintain a database of monthly workforce snapshots that track all available attributes of Canadian Armed Forces personnel. From this, monthly personnel counts for the populations separated by any attributes of interest can be easily extracted. The database also contains logs of attrition, promotions and transfers that associate these same attributes with each flowing member. From these logs, the number of flowing personnel

**Table 5.** Maximum absolute difference between direct application of the general formula and approximations derived from monthly data.

|                      | $\alpha_{MI}$ | $\alpha_{HI}$ | $\alpha_{UT}$ | $\alpha_{MC}$ |
|----------------------|-----------|-----------|-----------|-----------|
| Attrition            | 0.132446% | 0.124344% | 0.124349% | 0.124288% |
| Medical release      | 0.053305% | 0.050049% | 0.050044% | 0.050046% |
| Component transfer   | 0.034026% | 0.033708% | 0.033647% | 0.033680% |
| Promotion            | 0.116764% | 0.108619% | 0.108330% | 0.108583% |
| Occupation transfer  | 0.251360% | 0.241489% | 0.240426% | 0.241214% |
| Overall              | 0.251360% | 0.241489% | 0.240426% | 0.241214% |

from the sub-populations of interest can also easily be extracted. With these counts, approximations of the general formula for personnel flow rates based on monthly data may be readily computed. Evaluating the general formula based on daily data would require either substantially more effort each time a rate is required, or maintaining a database of daily snapshots, the size of which would cause difficulties in terms of required processing time on ordinary hardware.

Although the uniform Taylor approximation slightly outperforms the others with monthly data in Table 4 and Table 5, we find that it is less intuitive and harder to explain to our clients than the others. Considering the tradeoff between accuracy and simplicity, we have chosen to continue to prefer the half-intake formula for reporting personnel flow rates in the Canadian Armed Forces.

## 6    Potential Pitfalls of Rate Comparisons

Personnel flow rates are commonly used to compare populations. They offer a way of scaling flows in order to make them comparable between populations. However, while comparing rates is often instructive, it does not always provide the full story. In this section, we provide an example to illustrate the potential pitfall from relying exclusively on the juxtaposition of rates in the context of an employment equity assessment. This example is drawn from a Canadian Armed Forces gender study that was previously described in an internal report [14].

That study concerned Naval Technical Officers (NTOs) – an occupational category comprising Royal Canadian Navy engineers. It aimed to identify challenges faced by women through comparison to men in the same occupation, based on HR data covering 2004 to 2020. Among the statistics compiled for the study, it was found that, at the rank of Lieutenant Commanders (LCdr), women NTOs had experienced a 3.0% annual promotion rate (to the rank of Commander (Cdr)) over the period, versus 3.9% for men. At first glance, this is a substantial and troubling difference.

The potential pitfall, here, is that this rate comparison ignores other simultaneous flows. In this case, the important related flow is attrition. When an NTO departs, if everything else remains equal, the impact on the promotion rate is to increase it (by diminishing the denominator). Thus, the observed lower promotion rate of women could be due to women having experienced proportionally fewer promotions (possibly due to inequity), but could also be due to higher retention among women.

For this study, we obtained a fuller picture by tracking outcomes over time using an approach inspired from survival analysis [15]. We considered all of the NTOs who arrived at the rank of LCdr over the period (usually through promotion) and tracked what happened to them. They either ended up promoted to Cdr, released from the Canadian Armed Forces, or remained at the same rank at the end of the period. In Fig. 7, we show the proportion having either been promoted or released as a function of *time in rank*. To draw these graphs, the proportion is taken as the number promoted (respectively released) by a given time in rank, divided by the number of all those who would eventually be promoted or released, or who remained at the end of the period and had been a LCdr at least as long as the given time in rank.

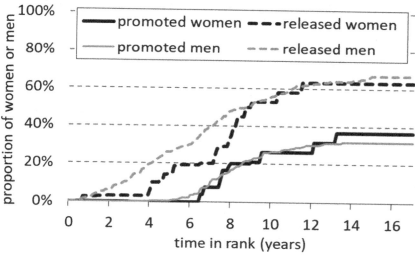

**Fig. 7.** Outcome experienced by LCdr NTOs over 2004–2020, as a function of time in rank.

Figure 7 contrasts the likely conclusion that would have been drawn from looking only at promotion rates. The proportion of women promoted actually follows closely that of men, even ending up a little higher. The main difference between women and men in Fig. 7 is rather that there have been proportionally much fewer releases of women before approximately eight years of time in rank. This is the difference that directly explains the difference in promotion rates. It is not to say that this situation is equitable (the fewer remaining men still did get promoted at a higher rate), but adds nuance to the interpretation of the promotion rates. We believe that comparing rates derived from the general formula is a valuable tool for Personnel OR practitioners, but that further investigation is warranted before decisions are made on the basis of such comparisons.

## 7 Conclusion

In this chapter, we have described and justified a general formula for measuring and reporting personnel flow rates. We believe that formally introducing this theoretically sound definition solidifies the foundation of Personnel OR as a discipline, and that it will increase the reliability and consistency of the figures being reported in practice. To ease the task of measuring these rates when the available data is not conducive to applying the general formula, we also introduced approximation formulas and empirically compared their accuracy. In light of our results, we put forward the half-intake approximation as our preferred approximation formula.

Personnel OR is not the only field where flow rates are commonly reported, compared and used for naïve forecasting. We saw that investment performance measurement is another field where rates of flow have been treated rigorously. Other areas, such as the reporting of churn rates for subscription services, might benefit from additional rigor and standardization in their treatment of proportional rates. We hope that our work in Personnel OR might inspire investigations in such areas.

Lastly, all of the empirical tests compiled for this chapter relied on Canadian Armed Forces personnel data – a large workforce for which much data was available to us, but one that might not be representative of other workforces. HR data is generally not shared externally, because of privacy considerations. In order to know if the same rate approximation formulas would behave similarly when applied to other organizations we would like to invite our Personnel OR colleagues to replicate our tests with the data available to them, so as to extend our results.

## References

1. Vincent, E., Okazawa, S., Calitoiu, D.: Attrition, promotion, transfer: reporting rates in personnel operations research. In: 10th International Conference on Operations Research and Enterprise Systems, pp. 115–122. SciTePress, Virtual (2021)
2. Noble, S.: Defining churn rate (no really, this actually requires an entire blog post), shopify engineering blog. https://shopify.engineering/defining-churn-rate-no-really-this-actually-equires-an-entire-blog-post. Accessed 22 Sept 2021
3. Okazawa, S.: Measuring attrition rates and forecasting attrition volume (DRDC CORA Technical Memorandum 2007-02). Defence Research and Development Canada, Ottawa (2007)
4. Bartholomew, D.J., Forbes, A.F., McClean, S.I.: Statistical Techniques for Manpower Planning, 2nd edn. Wiley, Chichester (1991)
5. Statistics Canada: Age-standardized Rates. The Daily, Statistics Canada, Ottawa (2017)
6. Latchman, S., Straver, M., Tanner, L.: Workforce analytics – the importance of context. Presentation at the International Conference on Analytics Driven Solutions, Ottawa (2014)
7. Fang, M., Bender, P.: Identifying contributors to changes in attrition (RTO-MP-SAS-081-17). North Atlantic Treaty Organization Research and Technology Organisation, Brussels (2010)
8. AT&T: Q1 2020 AT&T Earnings – Financial and Operational Trends. AT&T, Dallas (2020)
9. Canadian Securities Administrators: Notice of Amendments to National Instrument 31-103. CSA, Montreal (2017)
10. Chartered Financial Analyst Institute: Global Investment Performance Standards (GIPS®) For Firms. CFA Institute, Charlottesville (2019)

11. Chartered Financial Analyst Institute: Guidance Statement on Calculation Methodology, Global Investment Performance Standards. CFA Institute, Charlottesville (2011)

12. Vincent, E., Calitoiu, D., Ueno, R.: Personnel attrition rate reporting (DGMPRA Scientific Report DRDC-RDDC-2018-R238). Defence Research and Development Canada, Ottawa (2018)

13. Dietz, P.: Pension funds: measuring investment performance. Graduate School of Business, Columbia University, New York (1966)

14. Vincent, E.: Gender-focused analysis of Naval Technical Officers human resources data (DGMPRA Scientific Letter DRDC=RDDC-2021-L144). Defence Research and Development Canada, Ottawa (2021)

15. Miller, R.G.: Survival Analysis, 2nd edn. Wiley, New York (2011)

# Scenario-Based Optimization of Specific Group of Public Service Systems in Health Sector

Marek Kvet[✉] and Michal Kvet

Faculty of Management Science and Informatics, University of Žilina, Univerzitná 8215/1,
010 26 Žilina, Slovakia
{marek.kvet,michal.kvet}@fri.uniza.sk

**Abstract.** Service systems and their designing represent a very important field, in which the operations researchers and other specialists in optimization and Informatics find their irreplaceable application role. Situation becomes more serious in such cases, where money is not the crucial quality criterion. The family of systems studied in this paper consists of systems, in which the decisions affect human health or even life. Obviously, the key role in the decision-making process is played by the application of advanced location science. The main goal of this research paper is to provide the readers with a short spectrum of basic mathematical models and their comparison. The emphasis is put on scenario-based optimization, which takes into account different conditions in the network, through which the service is provided. Theoretical explanation of suggested methods is accompanied with the computational study performed with real-sized instances.

**Keywords:** Public service systems · Location science · Scenario-based optimization · Radial formulation

## 1 Introduction

This paper deals with specific applications of the location science. People have always analyzed how to locate facilities over a large area effectively. The term "facility" refers to various entities such as ports, factories, warehouses, retail outlets, but also schools, hospitals, bus stops, or even emergency warning sirens. The need for decisions on how to locate these facilities has led to a strong interest in operational research analysis and modeling. Using a mathematical model of a location problem, we look for the optimal location of different facilities in a given area and the assignment of customers to these facilities in order to optimize a certain criterion. The most important question raised in the process of solving the location problem is the selection of a suitable optimization criterion or objective function. The formulation of an objective function depends largely on the ownership of the organization. Private sector facilities are often located to minimize costs or maximize profit in any form. In contrast, the purpose of locating public facilities is more difficult to achieve. For example, if the task were to deploy emergency medical service centers, a possible criterion would be to minimize the average distance or travel time to the scene. Another suitable criterion would be to minimize the maximal distance

© Springer Nature Switzerland AG 2022
G. H. Parlier et al. (Eds.): ICORES 2020/2021, CCIS 1623, pp. 128–151, 2022.
https://doi.org/10.1007/978-3-031-10725-2_7

that the ambulance must travel before reaching the patient. Different interpretations of the goal of maximizing public satisfaction have led to a large number of different types of location problems [4–9, 14, 15, 18, 27, 28].

Generally, location problems can be either continuous or discrete. In continuous tasks, facilities or special devices can be located virtually anywhere within a given region. In discrete location problems, a finite set of candidates for center locations is given. In this paper, we focus on special class of discrete location problems for intelligent public service systems designing. Private systems can be left out here.

The family of locations problems applicable in public sector could be divided into two basic categories - media-based problems and coverage tasks. Media-based problems locate facilities so that the weighted average distance (or time) between customers and their associated center is minimal. The nodes of the transport network in which the facilities are to be located are called the medians of the network. These tasks are also referred to as location-allocation tasks, as they determine not only location (placement) but also allocation (assignment) decisions. The weighted $p$-median problem and the fixed-cost location problem are the most important types of tasks in this category. They differ in the following fact. While the $p$-median problem is formulated as a task to locate just $p$ facilities and it does not take into account the different costs of placing them in different nodes, the fixed-cost location problem seeks to minimize the total cost of both opening centers and traveling customers to the resort [1].

However, solutions that minimize weighted travel distance (or time) can be unfair and force some residents to travel too far. This can make customers reluctant to travel to the service center. It has even been shown that the usage of a service center decreases sharply when the distance (or time) of the journey exceeds a certain critical value [28]. Therefore, it is reasonable to take into account the maximal distance (or time limit) when solving the problem. This leads to the formulation of covering problems. Covering problems require locating the facilities so that they are able to serve each customer within a certain time (or distance). This category contains three main types of problems:

1. a location-covering problem in order to cover all customers with a minimal number of centers,
2. a location problem with maximal coverage to cover the maximal number of customers with a limited number of facilities,
3. a $p$-center problem in which it is necessary to find the location of $p$ facilities and all requirements must be covered and the maximal distance of the facility from the customer is minimal.

The location of service centers, special devices or any other facilities is therefore crucial both in industry and in services. If too many (or too few) facilities are open, this results in increased capital expenditures (or deterioration of services). Even the right number of improperly located devices unnecessarily reduces the quality of customer service. In healthcare, however, the consequences of bad decisions go beyond increased costs or deteriorating services. Improperly located rescue centers can have an impact on increased mortality or morbidity. Location science and optimization thus become much more important in the locations of medical facilities [11, 21, 25, 30, 32]. Obviously, the set of fields, in which the results of particular research can be applied is much wider and contains other public service systems [19, 27], intelligent manufacturing systems, traffic solutions, and many other areas.

In this research paper, we focus on the system of providing first aid in Slovakia, where we know the current state and the large amount of real data is available. Our goal is to shorten the travel time of Emergency Medical Service (EMS) stations and improve the availability of emergency care in Slovakia with different possible failures on the associated road network.

The remainder of this paper is organized as follows. The next section is devoted to the short overview of common mathematical models, which can find their application in public service system designing. They are discussed and compared. At the end of this section, there is a short analysis of their disadvantages, which has lead to the scenario-based optimization. The third section contains the basics of system robustness, which can be achieved by scenario-based optimization. Since the exact model optimizes only the situation of the worst possible case, other approaches based on processing the scenarios need to be introduced. Therefore, we discuss three different modeling methods, which have been originally presented at the ICORES conference. The theoretical explanation of suggested approaches is experimentally verified in the section entitled "Computational study". There is an overview of performed numerical experiments and the section yields also a brief comparative analysis of service center deployments based on different values of parameters. The methods are studied also from the point of computational time requirements. Finally, the last section brings a summary of the obtained results and offers some ideas for future possible research in this scientific field.

For completeness, let us focus now on the description of novelties brought by the extended version of the original manuscript [23]. The biggest difference between these two papers consists in the content and structure of this paper. While the original conference paper aims directly on three modelling approaches based on multi-objective principles, here we provide the readers with a wider research background and we try to place the suggested strategies into a big family of basic optimization approaches applied to EMS designing. Therefore, we report here also a brief overview of basic models, which find their application in the health sector and their analysis may lead to the necessity to use scenario-based optimization. Furthermore, the problem of EMS system designing is referenced by much more literature in this paper. The second and very important extension of the original paper consists in the practical applications of suggested modeling strategies on real world datasets. The originally reported computational study has been significantly extended and much more results are studied in this paper in order to make conclusions that are more relevant. The conference paper contained only a small sample of the whole set of performed experiments. Here, we provide the readers with a detailed analysis of the obtained results and compare the performance characteristics based on different values of parameters. The last but not least difference consists in the length of the paper. While the conference paper is short, this extension brings a detailed report on the studied problem.

## 2 Overview of Common Mathematical Models for Public Service System Designing in Health Sector

The main content of this section consists in a short list of mathematical models and their explanation. Presented models represent core types of approaches applicable in the decision-making process aimed at public service system designing.

The authors of study reported in [9] classified location models applied in healthcare into three categories: models based on accessibility, adaptability, and availability. They also argued that the three classical models of facility location (location-coverage problem, maximal coverage one, and the $p$-median problem) create the basis of most models for solving location problems in health care systems. Other scientists and researchers [10, 19, 27] confirmed mentioned theory also.

By accessibility we mean the ability of patients to reach a medical facility or the ability of EMS providers to reach patients. Models based on availability are formulated to locate health centers according to current conditions. Data on demand, travel costs, distance or time are generally considered to be fixed and non-random. Models tend to ignore changing system conditions or short-term fluctuations in facility availability. Such models include the maximal coverage model or the weighted $p$-median model. With regard to the current legislation issued by the Ministry of Health of the Slovak Republic, which intends to cover 95% of the population within 15 min, the usage of such types of mathematical models in practice is supported.

Adaptability models are used to determine the location of such facilites that cannot be easily or even relocated when conditions change. These decisions must be robust given the uncertain conditions in the future. Scenario planning is often used to handle the uncertain future. A number of scenarios is usually created to capture possible situations and a solution is selected in such a way that it will work in all (or most) scenarios. These types of models tend to look at the world in the long term and are used, for example, when planning the location of hospitals. Obviously, these types of models are bigger and require much more memory and computational time.

On the other hand, availability models, address very short-term changes in condition caused by facility occupancy when patients need it. That is why these models are most used for emergency response systems, where the vehicle (ambulance) may be busy handling the request at the time of another emergency situation. In terms of approach, they are divided into deterministic and probabilistic. A simple way to deal with station occupancy at a given time is to cover municipalities multiple times. Such a procedure was chosen by Daskin and Stern in creating the Hierarchical Objective Set Covering (HOSC) model [10], which first minimizes the number of stations needed to cover all nodes and then selects among the alternative optimal solutions one that maximizes the total multiple coverage. The number of stations that are able to meet the node coverage requirement gives multiple node coverage. The total multiple coverage is thus the sum of these multiple coverage of the individual nodes. Hogan and ReVelle [9] formulated a similar model, which is called backup coverage. This model requires additional coverage of each municipality by another station, while the objective function includes backup coverage weighted by the size of the request in the municipality. The above mentioned adaptability models use a deterministic approach to increase the availability of the station.

Furthermore, various probabilistic models based on the theory of collective service or on Bernoulli's experiments have been developed.

Let us now focus on the formulations of concrete mathematical models, which form the basic set of approaches for service system designing. We will move from general terminology to the terminology for the application area. The facilities in our case represent EMS stations, from which the ambulance vehicles are sent to solve the emergency at the scene. All municipalities in the Slovak Republic represent customers or generally system users. The number of inhabitants gives their weights.

One of the best-known models, with which we can solve the location problem to get the optimal EMS stations deployment in a large geographical area, is the weighted $p$-median problem. It tries to locate $p$ service centers ($p$ is a positive integer value) in such a way that it minimizes the total transport costs weighted by customer requirements [18]. It must be noted that the number of serviced customers as well as the number of possible service center locations may take the value of several thousand [3, 12]. The number of possible service center locations seriously impacts the computational time. Therefore, it is necessary to formulate a mathematical model in a proper way to achieve the resulting solution quickly enough.

To describe the above-mentioned problem by means of mathematical programming, let the symbol $I$ denote the set of possible service providing facility locations. Here, it is assumed that all candidates have equal setup cost and sufficient capacity to satisfy all assigned demands. Furthermore, let $J$ define the set of service recipients' locations, which contains various separate network points. The number of individual users located at $j \in J$ is denoted as $b_j$. The disutility for a user at the location $j$ following from the center location $i$ is denoted as non-negative $d_{ij}$. This problem has been broadly discussed in [3, 13, 16, 22] from the viewpoint of solving techniques for huge instances.

The basic decisions of the weighted $p$-median problem concern the location of service providing facilities at the network nodes from the set $I$ in order to minimize the sum of users' distances from their nearest facilities in such a way that the number of located centers does not exceed given value of $p$. In some models, exactly $p$ service centers are to be selected. Thus, the expression takes the form of equality. To model this decision at particular location, we introduce a zero-one variable $y_i \in \{0, 1\}$, which takes the value of 1, if a facility should be located at the candidate location $i$, and it takes the value of 0 otherwise. In addition, the allocation variables $z_{ij} \in \{0, 1\}$ for each $I \in I$ and $j \in J$ are introduced to assign a customer located at $j$ to a facility location $i$ by the value of 1. To meet the model conditions, the decision variables $y_i$ and also the allocation variables $z_{ij}$ have to satisfy the following constraints. Thus, the location-allocation model can be formulated by the expressions (1)–(6).

$$\text{Minimize} \quad \sum_{i \in I} \sum_{j \in J} b_j d_{ij} z_{ij} \tag{1}$$

$$\text{Subject to:} \quad \sum_{i \in I} z_{ij} = 1 \quad \text{for } j \in J \tag{2}$$

$$\sum_{i \in I} y_i = p \tag{3}$$

$$z_{ij} \leq y_i \quad \text{for } i \in I, j \in J \tag{4}$$

$$y_i \in \{0, 1\} \quad for \ i \in I \tag{5}$$

$$z_{ij} \in \{0, 1\} \quad for \ i \in I, j \in J \tag{6}$$

In the above model, the quality criterion (1) minimizes the sum of distances between customers and their nearest located sources of service. The allocation constraints (2) ensure that each client is assigned to exactly one facility location. Link-up constraints (4) enable to assign a customer $j$ to a possible center location $i$ only in such a case if the service center is located at this location and the formula (3) bounds the number of located centers by $p$ or it guarantees the exact number $p$ of located facilities. The problem (1)–(6) can be rewritten to a form acceptable by a modeler of integrated optimization software and solved by the associated IP-solver to optimality.

This approach may lead to an unfair resulting system design. Even if the average distance is minimal, the distance of the worst situated users to the nearest located facility may be very high. Consider the following example: If an emergency medical service system is designed, there is usually some period, within which the first aid must be provided and it makes sense. Otherwise, the rescuers' effort may be in vain. If an ambulance comes to a patient with stroke too late, the possibility of surviving rapidly decreases. Therefore, it is necessary to take into account mainly those users, whose system distance from the nearest located facility is higher. Here we suggest a modification of the objective function (1), where the distance values $d_{ij}$ are replaced by coefficients $c_{ij}$ defined as follows. Let $D_{max}$ represent the mentioned limit (time, distance or other quantitative value). If the disutility $d_{ij}$ is lower than or equal to $D_{max}$, then the coefficient $c_{ij}$ equals to 0, otherwise we set $c_{ij}$ to the former value of $d_{ij}$. Then the modified objective function takes the form of (7). Here the quality criterion combines the requirements of effectiveness and fairness together.

$$Minimize \quad \sum_{i \in I} \sum_{j \in J} b_j c_{ij} z_{ij} \tag{7}$$

The associated solving algorithm tries to cover all systems users by centers located near enough (particular distances do not exceed $D_{max}$) and if it is not possible, then the centers are located as near as possible. Another possibility to achieve certain level of fairness consists in consideration of the worst situated users instead of all of them.

Another basic way to decide on the location of EMS stations is to use the Location Set Covering Model (LSCM), first introduced by Toregas and colleagues in 1971. Its aim is to find the minimum number of facilities needed to cover all requirements [14].

As in the previous model, two sets $I$ and $J$ are given. Set $I$ represents the candidates for the location of the stations and set $J$ indicates all the municipalities to be covered. In addition, a maximal value $D_{max}$ must be defined. In order for a municipality $j$ to be covered by station $i$, the travel time or distance from this station to that municipality must be less than or equal to $D_{max}$. The set of all nodes that satisfy this condition for the municipality $j$ is denoted as the surroundings of the municipality $N_j = \{i \in I, d_{ij} \leq D_{max}\}$ for $j \in J$. For each node $i \in I$, a bivalent decision variable $y_i$ is defined, which assumes a value equal to 1 only if the station is located at the node $i$.

$$Minimize \quad \sum_{i \in I} y_i \tag{8}$$

$$Subject\ to: \quad \sum_{i \in Nj} y_i \geq 1 \quad for\ j \in J \tag{9}$$

$$y_i \in \{0,\ 1\} \quad for\ i \in I \tag{10}$$

The objective function (8) represents the number of located stations that needs to be minimized. The constraint (9) ensures that each municipality is covered by provided service – at least located center lies within radius $D_{max}$. The left part of the expressions represents the number of located centers that cover the municipality $j$. The last expression (10) is an obligatory constraint. However, according to [14], this model contains at least two significant shortcomings associated with its real application in practice:

1. The model cannot distinguish between municipalities that generate a relatively large and, on the contrary, relatively small demand at a given point in time.
2. The number of facilities needed to cover all requirements is often too large. However, it turned out that this is not the case in Slovakia.

In case we are not able to cover all requirements due to high costs, it is obvious that we will prefer to cover municipalities with higher demands. This idea has led to the publication [5] by Church and ReVelle, who formulated a coverage maximization model. This model, unlike the previous one, requires two additional inputs: The number of inhabitants or potential patients sharing the location $j$ denoted by $b_j$ and $p$ as the number of EMS stations to be located. Furthermore, we need extra variable $z_j$ for each location $j$. This variable will contain the information whether the location $j$ is covered within given limit or not. After these preliminaries, the mathematical model can be formulated as follows.

$$Maximize \quad \sum_{j \in J} b_j z_j \tag{11}$$

$$Subject\ to: \quad \sum_{i \in Nj} y_i \geq z_j \quad for\ j \in J \tag{12}$$

$$\sum_{i \in I} y_i = p \tag{13}$$

$$y_i \in \{0,\ 1\} \quad for\ i \in I \tag{14}$$

$$z_j \in \{0,\ 1\} \quad for\ j \in J \tag{15}$$

The objective function (11) maximizes the number of potential patients covered by the service. Constraint (12) specifies that a municipality $j$ cannot be marked as covered unless at least one station is located in its vicinity that would be able to cover it. The constraint (13) determines the exact number of stations to be located. The last two expressions (14) and (15) are obligatory constraints.

This model does not guarantee coverage of all municipalities. Furthermore, the fact that the village will be covered by one station does not necessarily mean that it will be served. In other words, the model at the strategic level does not take into account the capacity (or availability) of the facility, which may be used by another customer at the time the request arises. Therefore, it is necessary to extend this model to include aspects within the tactical level.

Let us focus now on a short evaluation of presented basic modeling approaches. Previous models do not take into account the need for a backup station to be sent to the patient if the only ambulance in the area is occupied. Therefore, the authors of Gendreau and colleagues presented the Double Standard Model (DSM), which includes double coverage of municipalities [14].

All presented mathematical models represent a core collection of possible approaches, which can be applied when a public service system should be designed or optimized. Obviously, there are many other models that combine presented ideas, i.e. min-max model, which minimizes the distance of the worst situated users from the nearest center, lexicographical model guaranteeing certain level of fairness in access to the service and many others.

As far as the solvability of practical instances from real world is concerned, the location-allocation formulation may bring several difficulties. The most serious one consists in the fact that the model contains too many decision variables and too many constraints for any common optimization environment. This obstacle can be overcome by so-called radial formulation, which was originally published in [12, 16, 21, 22]. The main idea of the radial approach consists in avoiding the allocation variables and in making the model much smaller. Let us explain the radial formulation on the first model (1)–(6). For each user location $j$, there is only one distance relevant, i.e. that is the distance to the nearest located center and this value affects the resulting objective function value. Since there is only one distance relevant for each location $j$ (see the constraint (2)), the allocation variables are not necessary anymore and this distance can be expressed using the following approach.

Let the symbol $m$ denote the maximal distance in the matrix $\{d_{ij}\}$, i.e. $m = \max\{d_{ij}: i \in I, j \in J\}$. In this computational study, we assume that all values in the matrix are integer. Of course, the radial model can be easily adjusted for real values. For each clients' location $j \in J$ and for each integer value $v = 0, 1 \ldots m\text{-}1$ we introduce a binary variable $x_{jv} \in \{0, 1\}$, which takes the value of one, if the distance $d_{j*}$ from the client located at $j \in J$ to the nearest EMS station is greater than the value of $v$ and it takes the value of zero otherwise. Then, the expression (16) holds for each $j \in J$.

$$d_{j*} = \sum_{v=0}^{m-1} x_{jv} \tag{16}$$

Similarly to the set-covering problem, a binary matrix $\{a^s_{ij}\}$ must be computed according to the formula (17).

$$a^v_{ij} = \begin{cases} 1 \text{ if } d_{ij} \le v \\ 0 \text{ otherwise} \end{cases} \quad for\ i \in I, j \in J, v = 0, 1, \ldots m - 1 \tag{17}$$

After these preliminaries, the radial model of the weighted $p$-median problem for obtaining the EMS stations deployment can be formulated by the expressions (18)–(22).

$$Minimize \quad \sum_{j \in J} b_j \sum_{v=0}^{m-1} x_{jv} \tag{18}$$

$$Subject \ to: \quad x_{jv} + \sum_{i \in I} a_{ij}^v y_i \geq 1 \quad for \ j \in J, \ v = 0, \ 1, \ \ldots, \ m-1 \tag{19}$$

$$\sum_{i \in I} y_i = p \tag{20}$$

$$y_i \in \{0, \ 1\} \quad for \ i \in I \tag{21}$$

$$x_{jv} \in \{0, \ 1\} \quad for \ j \in J, \ v = 0, \ 1, \ \ldots, \ m-1 \tag{22}$$

The quality criterion of the design formulated by the objective function (18) expresses the sum of distances from all clients to their nearest EMS station. The link-up constraints (19) ensure that the variables $x_{jv}$ are allowed to take the value of 0, if there is at least one center located in radius $v$ from the location $j$ and the constraint (20) limits the number of located stations by $p$. The last two series of obligatory constraints (21) and (22) keep the domain of the decision variables $y_i$ and $x_{jv}$.

The radial formulation enables us to solve much bigger instances than the former location-allocation approach [16, 21, 22]. That is why we will restrict ourselves only on the radial formulation of all mathematical models, which will be suggested and discussed in the remaining part of this paper.

Another very serious and big disadvantage of presented models consists in the fact, that they do not model reality in a sufficient level. It means that they assume some standard conditions in the network, through which the service is provided. But it must be realized that the decisions made by responsible authorities resulting from the optimization process are expected to affect the designed system especially in a healthcare segment for a long period of time. Therefore, selection of a suitable mathematical modelling tool has a strategic importance.

Consider, if a life of a person directly depends on the time, by which the first aid is provided, than the decision must be made in such a way, that the associated mathematical model takes into account as much real world features and as many possible side effects that may impact the quality of provided service. Here is a small example: Regardless the case whether the clients travel for service or the associated service is delivered to them, a transportation network must be used as a connection between service providers and their clients. Obviously, the traversing time between a service center and the affected user might be impacted by various random events caused by weather or traffic, and possible failure of a part of critical infrastructure should be taken into account. In other words, the system resistance to such critical events should be included into the decision-making process. Otherwise, the optimal solution does not have to fit real world situation, because the optimization process follows from some ideal conditions. Most of available approaches to increasing the system resistance [29, 31–33] are based on making the design resistant to possible failure scenarios, which can appear in the associated transportation network as a consequence of random failures due to congestion, disruptions or blockages. This way, a finite set of possible failure scenarios is usually given by the submitter of the optimization task in order to find such a solution that would fit to all scenarios to some extent. In other words, the scenario-based optimization plays a very important role in such cases, in which the decisions are expected to be implemented for long time.

## 3  Scenario-Based Optimization

Common simple optimization approaches discussed and presented in previous section do not hold to optimize public service systems mainly from the viewpoint of uncertainty and complexity of traffic, which may negatively affect the accessibility of service for clients in direct danger. Mainly in the healthcare segment, the situation becomes all the more serious that it can be about human lives and health.

If we want to make the resulting system design resistant to different adverse events, then the set of detrimental scenarios must be defined. Each scenario represents a possible change in the distance matrix $\{d_{ij}\}$, in which some elements give a higher value. Thus, each scenario brings a separate distance matrix. Logically, the number of scenarios must be kept as small as possible, because the number of possible service center locations may take the value of several thousands and the number of system users may take this value as well. Thus, each additional matrix is very memory demanding. If the scenario set contains too many scenarios, then the resulting mathematical model takes an unsolvable extent for almost all available solvers.

The standard approach to public service system is usually easy to handle due to a simple model structure, in which the average system accessibility for users (average response time) is often minimized. Unfortunately, the robust system design gets more complicated structure and the problem is formulated as a min-max model bringing some difficulties into the computational process [23]. Let us now formulate the most frequently referred model for robust service system designing.

As above, let symbol $J$ denote the set of users' locations and let symbol $I$ denote the set of possible service center locations. We denote by $b_j$ the number of users sharing the location $j$. In some cases, the number $b_j$ could express directly the estimated number of demands for service, but it is generally assumed that the number of ambulance vehicle arrivals to the location $j$ is proportional to the population density in this location. To obtain the result, $p$ locations must be chosen from $I$ so that the maximal scenario objective function value is to be the lowest possible. The objective function value of an individual scenario is computed as a sum of users' distances from the nearest facility multiplied by $b_j$. Furthermore, let symbol $U$ denote the set of possible failure scenarios. This set contains also one scenario called *basic scenario*, which represents standard conditions in the associated transportation network. The larger the set, the more difficult it is to find a solution to the problem. The integer distance between locations $i$ and $j$ under a specific scenario $u \in U$ is denoted by $d_{iju}$. Even if the radial model is originally suggested for integer distance or time values only, the used principle enables us to adjust the model also for real values.

The decision about locating a service center at any location $i$ from $I$ is modeled by decision variable $y_i$ for $i \in I$. It takes the value of 1 if a service center is located at $i$ and it equals to 0 otherwise. In the robust problem formulation, the variable $h$ as the upper bound of the objective function issues following the individual scenarios is used. To formulate the radial model, the integer range $[0, v]$ of all possible distances of the matrices $\{d_{iju}\}$ is partitioned into zones according to [16, 22]. The value of $v$ is computed according to the expression (23).

$$v = \max\{d_{iju} : i \in I, j \in J, u \in U\} - 1 \qquad (23)$$

To explain the used denotation, the zone $s$ corresponds to the interval $(s, s + 1]$. Furthermore, additional zero-one variables $x_{jus}$ need to be introduced for each triple $[j, u, s]$, where $j \in J$, $u \in U$ and $s = 0 \dots v$. The variable $x_{jus}$ equals to 1, if the distance $d_{ju*}$ of the client located at $j \in J$ from the nearest located facility in the scenario $u \in U$ is greater than $s$ and it takes the value of 0 otherwise. Then, the distance $d_{ju*}$ can be expressed as follows: $d_{ju*} = x_{ju0} + x_{ju1} + x_{ju2} + \dots + x_{juv}$. Similarly to the set covering problem, let us introduce a zero-one constant $a_{iju}^s$ for each scenario $u \in U$ and each $i \in I, j \in J$, $s \in [0..v]$. The constant $a_{iju}^s$ is equal to 1, if the distance $d_{iju}$ from the location $j$ to the candidate location $i$ is less than or equal to $s$, otherwise $a_{iju}^s$ is equal to 0. Then the mathematical model for obtaining the resulting system design takes the following form.

$$Minimize \quad h \qquad (24)$$

$$Subject\ to: \quad x_{jus} + \sum_{i \in I} a_{iju}^s y_i \geq 1 \quad for\ j \in J,\ u \in U,\ s = 0,\ 1,\ \dots,\ v \qquad (25)$$

$$\sum_{i \in I} y_i = p \qquad (26)$$

$$\sum_{j \in J} b_j \sum_{s=0}^{v} x_{jus} \leq h \quad for\ u \in U \qquad (27)$$

$$y_i \in \{0, 1\} \quad for\ i \in I \qquad (28)$$

$$x_{jus} \in \{0, 1\} \quad for\ j \in J,\ u \in U,\ s = 0,\ 1,\ \dots,\ v \qquad (29)$$

$$h \geq 0 \qquad (30)$$

To explain the above-formulated mathematical model, its objective function (24) gives an upper bound of all objective functions corresponding to the individual scenarios. The constraints (25) ensure that the variables $x_{jus}$ are allowed to take the value of 0, if there is at least one center located in radius $s$ from the user location $j$ and constraint (26) limits the number of located service centers by $p$. The link-up constraints (27) ensure that each scenario objective function is less than or equal to the upper bound $h$. The obligatory constraints (28), (29) and (30) are included to ensure the domain of the decision variables $y_i$, $x_{jus}$ and $h$.

Many authors [2, 26, 29, 31, 33, 34] have studied the topic of scenario-based optimization in order to develop various exact, heuristic and metaheuristic approaches, which could overcome the complexity of the model (24)–(30).

From our point of view, the exact mathematical model for robust service system design suffers from one important inconvenience. Let us focus on the set $U$ of detrimental scenarios. If we were able to estimate the probabilities of particular scenarios, then we could create one statistical scenario and make the model much smaller. The problem is that we are not able to do it, because of the lack of data or inability to predict randomly occurring events. Furthermore, the model (24)–(30) optimizes the service accessibility for all clients even for the worst possible case (the worst scenario). However, we do not know which scenario models the reality in the best way. It may happen that the optimal solution of the exact model (24)–(30) could be improved, if only the correct scenario was taken into account in spite of the whole set of scenarios. Therefore, we have developed three different approaches, which take into account individual scenarios in a more precise way. Mentioned approaches were originally published in [23]. Let us now focus on them in more details.

The first approach is based on the idea of minimizing the objective function value for the *basic scenario* under the condition that the objective functions connected with the detrimental scenarios do not change too much from their optimal values, which can be computed independently without the whole set of scenarios. As shown in [12, 22], the weighted $p$-median problem is easily solvable making use of the radial formulation. To achieve mentioned goal, the following denotation needs to be introduced and it will be used in the remaining part of this paper. If the index $u$ is set to the value of zero, it means that the *basic scenario* is concerned. In other words, the matrix $\{d_{ij0}\}$ corresponds to the *basic scenario*. If all the scenario objective functions are to be taken into account in the form of separate constraints, the goal value $G(u)$ for each scenario $u \in U_0$ should be computed. The expression (31) shows the weighted $p$-median problem solved for each failure scenario. Remember, that the symbol $U_0$ denotes the set of detrimental scenarios without the *basic scenario*.

$$G(u) = \min\left\{\sum_{j \in J} b_j \min\{d_{iju} : i \in I_1, I_1 \subseteq I, |I_1| = p\}\right\} \tag{31}$$

After introducing necessary denotations, a non-negative parameter $\varepsilon$ can be introduced. Its aim is to express the maximal increase of the $u$-th scenario objective function $G(u)$. The parameter $\varepsilon$ can either take a given exact value or it can be expressed as some percentage of the objective function $G(u)$. Then, the model for robust service system design can be formulated by the following expressions (32)–(37).

$$\text{Minimize} \quad \sum_{j \in J} b_j \sum_{s=0}^{v} x_{js0} \tag{32}$$

$$\text{Subject to :} \quad x_{jsu} + \sum_{i \in I} a_{iju}^s y_i \geq 1 \quad \text{for } j \in J, \ s = 0, \ 1, \ \ldots, \ v, \ u \in U \tag{33}$$

$$\sum_{i \in I} y_i = p \tag{34}$$

$$\sum_{j \in J} b_j \sum_{s=0}^{v} x_{jsu} \leq G(u) + \varepsilon \quad for \ u \in U_0 \tag{35}$$

$$y_i \in \{0, \ 1\} \quad for \ i \in I \tag{36}$$

$$x_{jsu} \in \{0, \ 1\} \quad for \ j \in J, \ s = 0, \ 1, \ \ldots, \ v, \ u \in U \tag{37}$$

Since the mathematical model (32)–(37) has very similar structure as the original exact model (24)–(30), it is not necessary to explain each constraint separately. As it was mentioned, the goal of this modeling approach is to optimize the objective function connected with the basic scenario only in such a way that the individual scenario objectives are not worsened too much.

The second suggested approach to robust public service system design follows from the previous method described by the model (32)–(37) and goal values $G(u)$ for all scenarios from the set $U_0$. In this approach, we replace the link-up constraints (35) of the original model by (38), in which only the maximal goal is taken into account.

$$\sum_{j \in J} b_j \sum_{s=0}^{v} x_{jsu} \leq MG + \varepsilon \quad for \ u \in U_0 \tag{38}$$

The maximal goal value denoted by $MG$ can be computed according to the following expression (39).

$$MG = \max\{G(u) : u \in U_0\} \tag{39}$$

The last, but not least important or interesting, modeling strategy follows from a different idea. On the contrary to the former principles, the third suggested approach brings the resulting service center deployment in such a way that the goal objective function value $G(0)$ for the *basic scenario* is computed first. After that, the value of parameter $\varepsilon$ must be given to limit the maximal possible increase of mentioned goal value $G(0)$. The objective function used in this approach minimizes possible increase $h$ of the maximal goal $MG$ over the set of scenarios. The associated mathematical model can be formulated in the following way.

$$Minimize \quad h \tag{40}$$

$$Subject \ to : \quad x_{jsu} + \sum_{i \in I} a_{iju}^s y_i \geq 1 \quad for \ j \in J, \ s = 0, \ 1, \ \ldots, \ v, \ u \in U \tag{41}$$

$$\sum_{i \in I} y_i = p \tag{42}$$

$$\sum_{j \in J} b_j \sum_{s=0}^{v} x_{jsu} \leq MG + h \quad for \ u \in U_0 \tag{43}$$

$$\sum_{j \in J} b_j \sum_{s=0}^{v} x_{js0} \leq G(0) + \varepsilon \tag{44}$$

$$y_i \in \{0, \ 1\} \quad for \ i \in I \tag{45}$$

$$x_{jsu} \in \{0, \ 1\} \quad for \ j \in J, \ s = 0, \ 1, \ \ldots, \ v, \ u \in U \tag{46}$$

$$h \geq 0 \tag{47}$$

The optimized quality criterion (40) expresses the increase of the maximal goal $MG$ over the scenarios objective functions. The formulae (41) ensure that the variables $x_{jus}$ are allowed to take the value of 0, if there is at least one facility located in radius $s$ from the user location $j$ and the expression (42) limits the number of located facilities by $p$. The link-up constraints (43) ensure that each individual scenario objective is less than or equal to the maximal goal value $MG$ increased by $h$. The constraint (44) does not allow to exceed given value of the objective function for the *basic scenario* $G(0)$ by more than $\varepsilon$. Finally, the obligatory constraints (45), (46) and (47) are included to ensure the domain of the decision variables.

## 4 Experimental Verification of Suggested Modeling Strategies

The main goal of this section is to study suggested modeling approaches from different points of view. First of all, if we want to decide on which of the presented models brings better system design, we have to have a tool for system design quality evaluation and not only by the value of optimized criterion, but also from the viewpoint of system robustness. Thus, a common comparative criterion for various modeling strategies needs to be introduced. It must be noted that each strategy optimizes different objective and therefore the comparative criterion should work with the resulting system design.

In accordance with previous sections, let $U$ denote the specific set of all scenarios containing also the *basic scenario*. Let $\mathbf{y}$ denote the vector of location variables $y_i$; $I \in I$. Let $\mathbf{y}^b$ correspond to the basic system design, i.e. the solution of a simple weighted $p$-median problem, in which only the *basic scenario* is taken into account. Let $f^b(\mathbf{y})$ denote the associated objective function value. Similarly, let $\mathbf{y}^r$ denote the solution of the robust model obtained by solving any of suggested models, which brings the robust system design. The robust objective $f^r(\mathbf{y})$ can be computed for any vector $\mathbf{y}$ of location variables according to the following expression (48).

$$f^r(\mathbf{y}) = \max \left\{ \sum_{j \in J} b_j \min \{ d_{iju} \ : \ y_i = 1 \} \ : \ u \in U \right\} \tag{48}$$

For completeness, (49) defines the original basic objective function value.

$$f^b(\mathbf{y}) = \sum_{j \in J} b_j \min \{ d_{ij0} \ : \ y_i = 1 \} \tag{49}$$

As far as the comparative criteria for different robust modeling strategies are concerned, let us introduce two coefficients.

The *price of robustness* (*POR*) expresses the relative increment (additional cost) of the basic scenario objective function, when $\mathbf{y}^r$ is applied instead of the optimal solution $\mathbf{y}^b$ obtained for the basic scenario. Its value is defined by (50).

$$POR = \frac{f^b(\mathbf{y}^r) - f^b(\mathbf{y}^b)}{f^b(\mathbf{y}^r)} \qquad (50)$$

The deficiency of the *price of robustness* consists in the fact that it does not express what we gain by such application of such solution. Therefore, we introduce also *gain of robustness* (*GOR*) according to (51).

$$GOR = \frac{f^r(\mathbf{y}^b) - f^r(\mathbf{y}^r)}{f^r(\mathbf{y}^r)} \qquad (51)$$

This coefficient evaluates the profit following from applying the robust solution instead of the standard one in the worst case ignoring the detrimental scenarios [23, 24].

After introducing the comparative criteria, we can analyze the suggested modeling strategies based on performed numerical experiments. All numerical experiments were performed using the optimization software FICO Xpress 7.3. They were run on a PC equipped with the Intel® Core™ i7 5500U processor with 2.4 GHz and 16 GB RAM.

As a source of benchmarks for this computational study, we used the data from real EMS system operated in eight self-governing regions of Slovakia. It means that the dataset used in our previous research reported in [23, 24] stays unchanged. The sizes of solved problems are reported in the following Table 1.

**Table 1.** Basic characteristics of used benchmarks.

| Region | BA | BB | KE | NR | TN | TT | ZA |
|--------|-----|-----|-----|-----|-----|-----|-----|
| $|I|$ | 87 | 515 | 460 | 350 | 276 | 249 | 315 |
| $p$ | 14 | 52 | 46 | 35 | 28 | 25 | 32 |

For each problem instance, ten different detrimental scenarios are available. More details about the process of scenario generating and about the used dataset can be found in many recent publications [8, 17, 20, 23, 24, 26].

The following tables summarize the results of the first modeling strategy described by the model (32)–(37) for various settings of parameter $\varepsilon$. This way, we have significantly extended the computational study from the original conference paper. The values reported in the tables correspond to basic studied characteristics: Let the symbol *ObjF* denote the particular model objective function. Computational time in seconds is reported in columns denoted by *CT* and finally, the coefficients *POR* and *GOR* are reported in percentage. Each table corresponds to one used benchmark.

**Table 2.** Results of the first modeling strategy described by the model (32)–(37) for the self-governing region of Bratislava (BA) and different values of parameter $\varepsilon$.

| $\varepsilon$ | ObjF | CT | POR | GOR |
|---|---|---|---|---|
| 1600 | Problem infeasible | | | |
| 1900 | Problem infeasible | | | |
| 3000 | 21999 | 125.97 | 8.15 | 32.18 |
| 11000 | 20342 | 11.34 | 0.00 | 0.00 |

**Table 3.** Results of the first modeling strategy described by the model (32)–(37) for the self-governing region of Banská Bystrica (BB) and different values of parameter $\varepsilon$.

| $\varepsilon$ | ObjF | CT | POR | GOR |
|---|---|---|---|---|
| 1600 | 17289 | 773.84 | 0.00 | 0.00 |
| 1700 | 17289 | 623.2 | 0.00 | 0.00 |
| 1800 | 17289 | 704.95 | 0.00 | 0.00 |
| 1900 | 17289 | 755.37 | 0.00 | 0.00 |
| 2000 | 17289 | 609.5 | 0.00 | 0.00 |
| 2100 | 17289 | 568.24 | 0.00 | 0.00 |
| 3000 | 17289 | 492.31 | 0.00 | 0.00 |
| 11000 | 17289 | 188.93 | 0.00 | 0.00 |

**Table 4.** Results of the first modeling strategy described by the model (32)–(37) for the self-governing region of Košice (KE) and different values of parameter $\varepsilon$.

| $\varepsilon$ | ObjF | CT | POR | GOR |
|---|---|---|---|---|
| 1600 | 20074 | 1405.58 | 0.16 | 4.1 |
| 1700 | 20063 | 448.48 | 0.10 | 3.55 |
| 1800 | 20063 | 929.06 | 0.10 | 3.55 |
| 1900 | 20063 | 734.51 | 0.10 | 3.55 |
| 2000 | 20063 | 1228.42 | 0.10 | 3.55 |
| 2100 | 20061 | 1209.87 | 0.09 | 1.98 |
| 3000 | 20042 | 392.39 | 0.00 | 0.00 |
| 11000 | 20042 | 356.89 | 0.00 | 0.00 |

**Table 5.** Results of the first modeling strategy described by the model (32)–(37) for the self-governing region of Nitra (NR) and different values of parameter $\varepsilon$.

| $\varepsilon$ | ObjF | CT | POR | GOR |
|---|---|---|---|---|
| 1600 | 22756 | 1382.98 | 0.46 | 3.76 |
| 1700 | 22747 | 1063.14 | 0.42 | 2.81 |
| 1800 | 22728 | 1299.69 | 0.34 | 2.24 |
| 1900 | 22708 | 1701.43 | 0.25 | 1.88 |
| 2000 | 22694 | 873.98 | 0.19 | 1.78 |
| 2100 | 22687 | 1057.05 | 0.16 | 1.31 |
| 3000 | 22651 | 268.94 | 0.00 | 0.00 |
| 11000 | 22651 | 199.75 | 0.00 | 0.00 |

**Table 6.** Results of the first modeling strategy described by the model (32)–(37) for the self-governing region of Trenčín (TN) and different values of parameter $\varepsilon$.

| $\varepsilon$ | ObjF | CT | POR | GOR |
|---|---|---|---|---|
| 1600 | 15757 | 586.55 | 0.45 | 2.9 |
| 1700 | 15739 | 215.8 | 0.34 | 3.19 |
| 1800 | 15739 | 478.44 | 0.34 | 3.19 |
| 1900 | 15739 | 250.03 | 0.34 | 3.19 |
| 2000 | 15721 | 339.21 | 0.22 | 3.53 |
| 2100 | 15706 | 101.65 | 0.13 | 3.26 |
| 3000 | 15686 | 75.93 | 0.00 | 0.00 |
| 11000 | 15686 | 89.04 | 0.00 | 0.00 |

**Table 7.** Results of the first modeling strategy described by the model (32)–(37) for the self-governing region of Trnava (TT) and different values of parameter $\varepsilon$.

| $\varepsilon$ | ObjF | CT | POR | GOR |
|---|---|---|---|---|
| 1600 | 18966 | 1174.82 | 0.49 | 3.55 |
| 1700 | 18966 | 1417.6 | 0.49 | 3.55 |
| 1800 | 18966 | 1422.85 | 0.49 | 3.55 |
| 1900 | 18926 | 803.92 | 0.28 | 1.22 |
| 2000 | 18926 | 1333.08 | 0.28 | 1.22 |
| 2100 | 18886 | 535.6 | 0.07 | 0.93 |
| 3000 | 18873 | 133.74 | 0.00 | 0.00 |
| 11000 | 18873 | 54.34 | 0.00 | 0.00 |

**Table 8.** Results of the first modeling strategy described by the model (32)–(37) for the self-governing region of Žilina (ZA) and different values of parameter $\varepsilon$.

| $\varepsilon$ | ObjF | CT | POR | GOR |
|---|---|---|---|---|
| 1600 | 21320 | 1939.85 | 1.55 | 8.53 |
| 1700 | 21320 | 742.61 | 1.55 | 8.53 |
| 1800 | 21320 | 4593.59 | 1.55 | 8.53 |
| 1900 | 21238 | 2049.56 | 1.16 | 7.03 |
| 2000 | 21215 | 2069.36 | 1.05 | 6.51 |
| 2100 | 21205 | 2011.15 | 1.00 | 7.18 |
| 3000 | 21119 | 779.41 | 0.59 | 4.99 |
| 11000 | 20995 | 97.58 | 0.00 | 0.00 |

The results reported in Tables 2, 3, 4, 5, 6, 7 and 8 proved that the resulting solution of the first modeling strategy is strongly dependent on the setting of parameter $\varepsilon$. If this parameter is too high, then we obtain the same solution as by solving a simple weighted $p$-median problem, what can be easily explained. If we allow to increase the scenario objective function too much, then the robustness principle loses its sense. As far as the computational time is concerned, this may be dependent on the size of the model and influenced by the model structure as well. Another conclusion leading from the obtained results can be formulated in such a way, that the weaker the limitations for individual scenario objective function value (higher values of $\varepsilon$), the lower is the price for robustness. Let us now focus on the second strategy, which makes a simple adjustment by (38) and (39). The structure of the following tables keep the former rules applied in Tables 2, 3, 4, 5, 6, 7 and 8.

**Table 9.** Results of the second strategy adjusted by constraints (38) and (39) for the self-governing region of Bratislava (BA) and different values of parameter $\varepsilon$.

| $\varepsilon$ | ObjF | CT | POR | GOR |
|---|---|---|---|---|
| 1600 | 22050 | 138.19 | 8.4 | 36.24 |
| 1700 | 22050 | 121.3 | 8.4 | 36.24 |
| 1800 | 22008 | 107.4 | 8.19 | 32.58 |
| 1900 | 21999 | 113.46 | 8.15 | 32.18 |
| 2000 | 21999 | 152.77 | 8.15 | 32.18 |
| 2100 | 21999 | 115.35 | 8.15 | 32.18 |
| 3000 | 21999 | 75.73 | 8.15 | 32.18 |
| 11000 | 20342 | 9.47 | 0.00 | 0.00 |

**Table 10.** Results of the second strategy adjusted by constraints (38) and (39) for the self-governing region of Banská Bystrica (BB) and different values of parameter $\varepsilon$.

| $\varepsilon$ | ObjF | CT | POR | GOR |
|---|---|---|---|---|
| 1600 | 17289 | 710.08 | 0.00 | 0.00 |
| 1700 | 17289 | 694.69 | 0.00 | 0.00 |
| 1800 | 17289 | 737.44 | 0.00 | 0.00 |
| 1900 | 17289 | 734.07 | 0.00 | 0.00 |
| 2000 | 17289 | 655.2 | 0.00 | 0.00 |
| 2100 | 17289 | 639.15 | 0.00 | 0.00 |
| 3000 | 17289 | 788.44 | 0.00 | 0.00 |
| 11000 | 17289 | 191.15 | 0.00 | 0.00 |

**Table 11.** Results of the second strategy adjusted by constraints (38) and (39) for the self-governing region of Košice (KE) and different values of parameter $\varepsilon$.

| $\varepsilon$ | ObjF | CT | POR | GOR |
|---|---|---|---|---|
| 1600 | 20055 | 841.13 | 0.06 | 1.46 |
| 1700 | 20055 | 754.54 | 0.06 | 1.46 |
| 1800 | 20055 | 848.33 | 0.06 | 1.46 |
| 1900 | 20042 | 412.27 | 0.00 | 0.00 |
| 2000 | 20042 | 424.4 | 0.00 | 0.00 |
| 2100 | 20042 | 422.52 | 0.00 | 0.00 |
| 3000 | 20042 | 434.31 | 0.00 | 0.00 |
| 11000 | 20042 | 368.52 | 0.00 | 0.00 |

**Table 12.** Results of the second strategy adjusted by constraints (38) and (39) for the self-governing region of Nitra (NR) and different values of parameter $\varepsilon$.

| $\varepsilon$ | ObjF | CT | POR | GOR |
|---|---|---|---|---|
| 1600 | 22756 | 1132.67 | 0.46 | 3.76 |
| 1700 | 22747 | 1064.14 | 0.42 | 2.81 |
| 1800 | 22728 | 836.65 | 0.34 | 2.24 |
| 1900 | 22708 | 1369.7 | 0.25 | 1.88 |
| 2000 | 22694 | 1607.78 | 0.19 | 1.78 |
| 2100 | 22687 | 1023.53 | 0.16 | 1.31 |
| 3000 | 22651 | 250.01 | 0.00 | 0.00 |
| 11000 | 22651 | 167.85 | 0.00 | 0.00 |

**Table 13.** Results of the second strategy adjusted by constraints (38) and (39) for the self-governing region of Trenčín (TN) and different values of parameter $\varepsilon$.

| $\varepsilon$ | ObjF | CT | POR | GOR |
|------|-------|--------|------|------|
| 1600 | 15706 | 472.95 | 0.13 | 3.26 |
| 1700 | 15688 | 119.98 | 0.01 | 2.11 |
| 1800 | 15688 | 76.18 | 0.01 | 2.11 |
| 1900 | 15688 | 128.41 | 0.01 | 2.11 |
| 2000 | 15688 | 146.65 | 0.01 | 2.11 |
| 2100 | 15686 | 94.43 | 0.00 | 0.00 |
| 3000 | 15686 | 188.75 | 0.00 | 0.00 |
| 11000 | 15686 | 89.62 | 0.00 | 0.00 |

**Table 14.** Results of the second strategy adjusted by constraints (38) and (39) for the self-governing region of Trnava (TT) and different values of parameter $\varepsilon$.

| $\varepsilon$ | ObjF | CT | POR | GOR |
|------|-------|---------|------|------|
| 1600 | 18939 | 1011.53 | 0.35 | 1.72 |
| 1700 | 18926 | 1500.82 | 0.28 | 1.22 |
| 1800 | 18886 | 488.64 | 0.07 | 0.93 |
| 1900 | 18886 | 560.12 | 0.07 | 0.93 |
| 2000 | 18873 | 138.12 | 0.00 | 0.00 |
| 2100 | 18873 | 134.16 | 0.00 | 0.00 |
| 3000 | 18873 | 134.23 | 0.00 | 0.00 |
| 11000 | 18873 | 54.36 | 0.00 | 0.00 |

**Table 15.** Results of the second strategy adjusted by constraints (38) and (39) for the self-governing region of Žilina (ZA) and different values of parameter $\varepsilon$.

| $\varepsilon$ | ObjF | CT | POR | GOR |
|------|-------|---------|------|------|
| 1600 | 21320 | 539.38 | 1.55 | 8.53 |
| 1700 | 21320 | 653.15 | 1.55 | 8.53 |
| 1800 | 21320 | 587.45 | 1.55 | 8.53 |
| 1900 | 21205 | 1461.05 | 1.00 | 7.18 |
| 2000 | 21205 | 1595.62 | 1.00 | 7.18 |
| 2100 | 21205 | 1726.00 | 1.00 | 7.18 |
| 3000 | 21119 | 559.23 | 0.59 | 4.99 |
| 11000 | 20995 | 84.03 | 0.00 | 0.00 |

The achieved results reported in Tables 9, 10, 11, 12, 13, 14 and 15 have similar characteristics as those obtained by the first suggested modeling approach discussed above. The only difference consists in the goal objectives considered within individual scenarios. The optimization process of the second approach finished its search in short time in most cases.

The last portion of performed numerical experiments was focused on the third modeling strategy, which considers objective function values of individual scenarios according to the model (40)–(47). Here, it can be logically expected that the computational times will be in order higher due to the model structure and used min-max criterion to be minimized. Replacing the min-sum objective by a min-max criterion ruins previous good

**Table 16.** Results of the third strategy described by the model (40)–(47) for the self-governing regions of Slovakia and different values of parameter $\varepsilon$.

| Region | $\varepsilon$ | ObjF | CT | POR | GOR |
|--------|------|------|------|------|------|
| BA | 500 | 6871 | 83.67 | 1.70 | 10.98 |
| | 1000 | 4224 | 123.74 | 4.12 | 21.26 |
| | 1500 | 3217 | 104.41 | 6.88 | 25.70 |
| | 2000 | 1085 | 87.2 | 8.40 | 36.24 |
| BB | 500 | 386 | 18380.03 | 2.61 | 2.73 |
| | 1000 | 386 | 17377.83 | 2.61 | 2.73 |
| | 1500 | 386 | 18053.98 | 2.61 | 2.73 |
| | 2000 | 386 | 17506.12 | 2.61 | 2.73 |
| KE | 500 | 548 | 15768.54 | 2.16 | 6.29 |
| | 1000 | 548 | 21768.74 | 2.16 | 6.29 |
| | 1500 | 548 | 23663.18 | 2.16 | 6.29 |
| | 2000 | 548 | 29934.04 | 2.16 | 6.29 |
| NR | 500 | 710 | 15775.37 | 2.14 | 6.84 |
| | 1000 | 657 | 12695.91 | 2.47 | 7.07 |
| | 1500 | 657 | 13425.98 | 2.47 | 7.07 |
| | 2000 | 657 | 12269.29 | 2.47 | 7.07 |
| TN | 500 | 328 | 1522.57 | 3.16 | 9.85 |
| | 1000 | 284 | 2533.87 | 4.05 | 10.13 |
| | 1500 | 284 | 2378.13 | 4.05 | 10.13 |
| | 2000 | 284 | 2317.85 | 4.05 | 10.13 |
| TT | 500 | 998 | 2226.63 | 2.08 | 4.61 |
| | 1000 | 788 | 1821.63 | 3.87 | 5.68 |
| | 1500 | 788 | 2267.57 | 3.87 | 5.68 |
| | 2000 | 788 | 2037.03 | 3.87 | 5.68 |
| ZA | 500 | 601 | 2957.48 | 2.33 | 12.93 |
| | 1000 | 525 | 4799.91 | 3.75 | 13.31 |
| | 1500 | 525 | 4196.03 | 3.75 | 13.31 |
| | 2000 | 525 | 4293.32 | 3.75 | 13.31 |

solvability of the model even if the radial formulation is applied. The obtained results are summarized in the following Table 16.

Analyzing the results in Table 16, the expectations have been confirmed. The quality of obtained designs measured by the coefficients *POR* and *GOR* depend on the parameter settings. Based on reported results, presented approaches represent suitable contribution to the state-of-the-art methods for robust service system designing.

## 5  Conclusions

This paper was aimed at the problem of public service system designing making use of means of mathematical modeling and operations research methods.

The first part of the paper brought a short overview of common optimization approaches, which form the core collection of basic models applicable in practical life. Based on their analysis from the viewpoint of period in which the decisions affect the system performance, it was suggested to formulate a special set of scenarios describing possible failures and other unexpected events on the network. These detrimental events may negatively affect the efficiency of provided service. If the decision affects health or even life of the served population, as many different aspects of the real system as possible should be taken into account in the optimization process in order to make the final decision good and acceptable.

Based on this point of view, the concept of system robustness was introduced and various scenario-based optimization strategies have been developed. In this paper, we suggested and studied three different modeling strategies, in which the goal values of individual scenario objective function values were taken into account in the form of special constraints.                                                    .

The last, but not least important, part of the paper was focused on a computational study, in which several characteristics of suggested modeling approaches were experimentally verified and studied. It can be observed in the reported results that the computational time demands depend on the model structure. If we replace a min-sum objective by a min-max optimization criterion, then the model gets more complicated so it requires a longer computation time. Besides that, quality of obtained results was very satisfactory. Thus, we have constructed very useful approaches that are applicable for public service system designing mainly in such situations, in which system robustness must be taken into account.

Obviously, presented results have brought several challenges that need to be covered in the future. Attention could be paid to several new modeling and solving techniques, which could reach the resulting system design faster and keep satisfactory solution accuracy. Another challenge consists in mastering the problem size, which grows with increasing number of used detrimental scenarios. Finally, an additional possible research goal would be to create such an optimization method, which allows to locate more service centers to the same location under the condition of keeping the system robust enough.

**Acknowledgment.** This work was supported by the research grant VEGA 1/0216/21 "Design of emergency systems with conflicting criteria using artificial intelligence tools". This work was supported also by the Slovak Research and Development Agency under the Contract no. APVV-19-0441.

# References

1. Ahmadi-Javid, A., Seyedi, P., et al.: A survey of healthcare facility location. Comput. Oper. Res. **79**, 223–263 (2017). ISSN 0305-0548
2. Antunes, C.H. and Henriques, C.O.: Multi-objective optimization and multi-criteria analysis models and methods for problems in the energy sector. In: Greco, S., Ehrgott, M., Figueira, J. (eds.) Multiple Criteria Decision Analysis. International Series in Operations Research & Management Science, vol. 233, 1071–1169. Springer, New York (2016). https://doi.org/10.1007/978-1-4939-3094-4_25
3. Avella, P., Sassano, A., Vasil'ev, I.: Computational study of large scale p-median problems. Math. Program. **109**(1), 89–114 (2007)
4. Brotcorne, L., Laporte, G., Semet, F.: Ambulance location and relocation models. Eur. J. Oper. Res. **147**, 451–463 (2003)
5. Church, R.L., Murray, A.: Location Covering Models: History, Applications and Advancements. Springer, Cham (2018). 271s. ISBN 978-3-319-99845-9
6. Correia, I., Saldanha-da-Gama, F.: Facility location under uncertainty. In: Laporte, G., Nickel, S., Saldanha da Gama, F. (eds.) Location Science, pp. 185–213. Springer, Cham (2019). https://doi.org/10.1007/978-3-030-32177-2_8
7. Current, J., Daskin, M., Schilling, D.: Discrete network location models. In: Drezner, Z., et al. (eds.) Facility Location. Applications and Theory, pp. 81–118. Springer, Berlin (2002). https://link.springer.com/gp/book/9783540421726
8. Czimmermann, P., Koháni, M.: Characteristics of changes of transportation performance for pairs of critical edges. Commun. Sci. Lett. Univ. Žilina **20**(3), 84–87 (2018)
9. Daskin, M.S., Dean, L.K.: Location of health care facilities. In: Brandeau, M.L., Sainfort, F., Pierskalla, W.P. (eds.) Operations Research and Health Care. International Series in Operations Research & Management Science, vol. 70, pp. 43–76. Springer, Boston (2005). https://doi.org/10.1007/1-4020-8066-2_3
10. Daskin, M., Hogan, K., ReVelle, C.: Integration of multiple, excess, backup, and expected covering models. Environ. Plann. B. Plann. Des. **15**, 15–35 (1988). https://doi.org/10.1068/b150015
11. Doerner, K.F., et al.: Heuristic solution of an extended double-coverage ambulance location problem for Austria. Cent. Eur. J. Oper. Res. **13**(4), 325–340 (2005)
12. García, S., Labbé, M., Marín, A.: Solving large p-median problems with a radius formulation. INFORMS J. Comput. **23**(4), 546–556 (2011)
13. Gendreau, M., Potvin, J.: Handbook of Metaheuristics, 648 p. Springer, Heidelberg (2010)
14. Gunes, E., Melo, T., Nickel, S.: Location Problems in Healthcare, Electronic (2019). ISBN: 978-3-030-32177-2
15. Ingolfsson, A., Budge, S., Erkut, E.: Optimal ambulance location with random delays and travel times. Health Care Manag. Sci. **11**(3), 262–274 (2008)
16. Janáček, J.: Approximate covering models of location problems. In: Lecture Notes in Management Science: Proceedings of the 1st International Conference ICAOR, Yerevan, Armenia, pp. 53–61 (2008)
17. Janáček, J., Kvet, M.: Detrimental scenario construction based on network link characteristics. In: Proceedings of the 19th IEEE International Carpathian Control Conference (ICCC), Szilvásvárad, pp. 629–632 (2018)
18. Jánošíková, Ľ: Emergency Medical Service Planning. Commun. Sci. Lett. Univ. Žilina **9**(2), 64–68 (2007)
19. Jánošíková, Ľ, Jankovič, P., Kvet, M., Zajacová, F.: Coverage versus response time objectives in ambulance location. Int. J. Health Geogr. **20**, 1–16 (2021)

20. Jenelius, E.: Network structure and travel patterns: explaining the geographical disparities of road network vulnerability. J. Transp. Geogr. **17**, 234–244 (2009)
21. Kvet, M.: Computational study of radial approach to public service system design with generalized utility. In: Digital Technologies 2014: Proceedings of the 10th International IEEE Conference, pp. 198–208 (2014)
22. Kvet, M.: Advanced radial approach to resource location problems. In: Rocha, Á., Reis, L.P. (eds.) Developments and Advances in Intelligent Systems and Applications. SCI, vol. 718, pp. 29–48. Springer, Cham (2018). https://doi.org/10.1007/978-3-319-58965-7_3
23. Kvet, M.: Robust emergency medical system design as a multi-objective goal programming problem. In: ICORES 2021: Proceedings of the 10th International Conference on Operations Research and Enterprise Systems, pp. 21–28. SCITEPRESS (2021). ISBN 978-989-758-485-5, ISSN 2184-4372
24. Kvet, M., Janáček, J.: Robust emergency system design using reengineering approach. In: ICORES 2020: 9th International Conference on Operations Research and Enterprise Systems, Valletta, Malta, pp. 172–178 (2020)
25. Kvet, M., Janáček, J., Kvet, M.: Computational study of emergency service system reengineering under generalized disutility. In: Parlier, G.H., Liberatore, F., Demange, M. (eds.) ICORES 2018. CCIS, vol. 966, pp. 198–219. Springer, Cham (2019). https://doi.org/10.1007/978-3-030-16035-7_11
26. Majer, T., Palúch, S.: Rescue system resistance to failures in transport network. In: Proceedings of 34th International Conference Mathematical Methods in Economics, pp. 518–522. Technical University of Liberec, Liberec (2016)
27. Marianov, V., Serra, D.: Location problems in the public sector. In: Drezner, Z. (ed.) Facility Location - Applications and Theory, pp. 119–150. Springer, Berlin (2002). https://link.springer.com/gp/book/9783540421726
28. Rahman, S., Smith, D.: Use of location-allocation models in health service development planning in developing nations. Eur. J. Oper. Res. **123**, 437–452 (2000). https://doi.org/10.1016/S0377-2217(99)00289-1
29. Pan, Y., Du, Y., Wei, Z.: Reliable facility system design subject to edge failures. Am. J. Oper. Res. **4**, 164–172 (2014)
30. Reuter-Oppermann, M., van den Berg, P.L., Vile, J.L.: Logistics for emergency medical service systems. Health Syst. **6**(3), 187–208 (2017)
31. Scaparra, M.P., Church, R.L.: Location problems under disaster events. In: Laporte, G., Nickel, S., da Gama, F.S. (eds.) Location Science, pp. 623–642. Springer, Cham (2015). https://doi.org/10.1007/978-3-319-13111-5_24
32. Schneeberger, K., Doerner, K.F., Kurz, A., Schilde, M.: Ambulance location and relocation models in a crisis. Cent. Eur. J. Oper. Res. **24**(1), 1–27 (2014). https://doi.org/10.1007/s10100-014-0358-3
33. Snyder, L.V., Daskin, M.S.: Reliability models for facility location; The expected failure cost case. Transp. Sci. **39**(3), 400–416 (2005)
34. Sullivan, J.L., Novak, D.C., Aultman-Hall, L., Scott, D.M.: Identifying critical road segments and measuring system-wide robustness in transportation networks with isolating links: a link-based capacity-reduction problem. Transp. Res. Part A **44**, 323–336 (2010)

# Project Selection with Uncertainty Using Monte Carlo Simulation and Multi-criteria Decision Methods

Guilherme Augusto Barucke Marcondes$^{(\boxtimes)}$ and Marina da Silva Vilela

National Institute of Telecommunications, Inatel, Santa Rita do Sapucaí, Brazil
guilherme@inatel.br, marina.vilela@pg.inatel.br
http://www.inatel.br

**Abstract.** Project selection is a frequent challenge for companies. Usually, human, material and financial resources are scarce and not sufficient to carry out all the candidate. Additionally, decision-makers still need to deal with estimation uncertainty, which is inherent in projects, especially those in Research and Development (R&D). Multi-criteria decision methods (MCDM) can be important tools, allowing the evaluation through ranking, which considers several criteria, often in conflict with each other. This work proposes a method to incorporate uncertainty in project selection using together ELECTRE II, PROMETHEE II, VIKOR and TOPSIS methods, by means of Monte Carlo simulation. It is exemplified proceeding selection from a real set of information technology product development projects.

**Keywords:** Project selection · Multi-criteria decision methods · TOPSIS · VIKOR · ELECTRE II · PROMETHEE II · Uncertainty

## 1 Introduction

A common challenge in companies is the selection of projects. As they must share financial support, laboratories, equipment and specialized personnel, for instance, usually, resources are scarce and not enough to run them all at the same time [2,7,13]. Decision-makers must then select a subset of them from among those that are presented as candidates for execution [1]. There are several methods to support this decision, with ranking being a useful tool that allows a choice based on structured forms of comparison, especially if objective criteria are defined, with a view to greater alignment with the companies' market strategies and demands [25].

Project selection usually depends on comparing several criteria simultaneously. For helping in this task, Multi-criteria Decision Methods (MCDM) allow one to build a ranking, listing candidates from the best to the worst options [19,35]. The application of MCDM has grown in scientific literature [29].

Examples of MCDM are *Elimination Et Choix Traduisant la Réalité* II (ELEC-TRE II), Preference Ranking Organization Method for Enrichment Evaluation II (PROMETHEE II), *VIseKriterijumska Optimizacija I Kompromisno Resenje* (VIKOR),

© Springer Nature Switzerland AG 2022
G. H. Parlier et al. (Eds.): ICORES 2020/2021, CCIS 1623, pp. 152–170, 2022.
https://doi.org/10.1007/978-3-031-10725-2_8

and Technique for Order of Preference by Similarity to Ideal Solution (TOPSIS) [5, 10, 22, 28]. As an output, they all allow the construction of rankings supporting decision process [17].

In [36], the authors used PROMETHEE II as ranking tool for dealing with five criteria and 13 sub-criteria in projects of solar power plants. The authors of [31] applied an improvement of TOPSIS in risk evaluation of investing in projects. VIKOR is the method for risk analysis in new product projects, proposing a decision support tool in [21]. The paper [27] proposed an ELECTRE based method for project ranking, using it for selecting gas hydrate exploitation projects, evaluating social, environmental and economic criteria. A point to be noted is that the application of these four methods on the same selection of projects for decision can lead to different rankings.

Uncertainty is significant on the selection of Research and Development (R&D) projects, which can have a negative impact on a company's future if the results are not as expected [13]. Given that uncertainty is inherent in R&D a [19], companies should select them carefully to avoid wasting resources [34]. And the application of formal selection methods increases the chances of success [7].

Specialists must examine the criteria and estimate objective values for each project in order to use MCDM. It involves selection uncertainty, which is inherent in the estimation process and is a common and unavoidable occurrence in projects [4]. As a result, when using the MCDM, uncertainty should be taken into account.

Three-point estimation approach incorporates the variation in values produced by estimating, using three values instead of a single value: most likely, worst-case, and best-case [12, 26, 38]. A triangle probability distribution for the parameters under consideration can be built based on these three values, allowing to address uncertainty in evaluations [20, 26, 32, 37].

Monte Carlo simulation is a useful tool for taking into account uncertainty in judgment needed for projects [16, 19, 26]. Based on the defined probability distribution, each simulation round generates a random value for the parameters. The variation in the results might be seen after multiple rounds.

In [14], it was proposed an approach for ranking supporting project selection based on ELECTRE II method and considering uncertainty. It is a very useful tool, but it considers just one MCDM. In this work, an extended version of [14], we propose a way of considering uncertainty in ranking by MCDM, applying four methods: PROMETHEE II, ELECTRE II, TOPSIS and VIKOR. As they can lead to different rankings, because they apply different comparison techniques, we propose, additionally, a way to bring these methods together into a single prioritization list. For each project, three-point estimation is proceeded, triangular distributions are set and Monte Carlo simulation done. The proposed method is exemplified by applying over a set of ten real hardware and software development projects (from an information technology equipment provider company).

## 2    Multi-criteria Decision Methods

When the decision to choose among alternatives involves only one criterion, the selection is quite intuitive, as it is enough to select the ones that present the best scores.

However, when the choice involves more than one criterion, the decision can be more complex, as the criteria must be evaluated simultaneously and, in some cases, there may be a conflict among them [33].

Works in literature indicate an increase in the use of Multi-criteria Decision Methods (MCDM) in project selection [29]. These methods allow one to classify different alternatives considering various criteria and indicate the best ones.

Four MCDM found in the literature are: *ELimination Et Choix Traduisant la Réalité* II (ELECTRE II), Preference Ranking Organization Method for Enrichment Evaluation II (PROMETHEE II), *VIseKriterijumska Optimizacija I Kompromisno Resenje* (VIKOR), and Technique for Order of Preference by Similarity to Ideal Solution (TOPSIS). ELECTRE II allows one to build a preference classification by calculating agreement and disagreement indices, comparing all alternatives to each other [33]. PROMETHEE II is a method for ranking alternatives with finite decision criteria. TOPSIS searches for the best alternatives by identifying how close they are to the ideal solution and how far they are from the non-ideal solution. VIKOR evaluates the performance of alternatives with several comparison criteria. The closer to the ideal solution, the better the alternative [17].

For project selection, MCDM must follow the steps (adapted from [23]):

– Define evaluation criteria that are related to the intended goals;
– Evaluate the alternatives according to the defined criteria;
– Apply MCDM;
– Select one or more alternatives.

At the end, the MCDM delivers a list of prioritized alternatives (for this work, the alternatives are the projects that are candidates for execution). Based on this list, decision-maker may choose the best ranked one(s), depending on how many will be selected.

### 2.1  *Elimination Et Choix Traduisant la Réalité* II (ELECTRE II)

ELECTRE (in English, Elimination and Choice Translating Reality) methods have different approaches (as in I, II, III and IV). They have also different applications [24]:

– ELECTRE I for selection, but without a ranking;
– ELECTRE II, III and IV for ranking problems;
– ELECTRE II and III when it is possible and desirable to quantify the relative importance;
– ELECTRE III incorporates the fuzzy nature of decision-making;
– ELECTRE IV when quantification is not possible.

For this work, the chosen method was ELECTRE II, due to the possibility of ranking the alternatives and because it is possible quantifying relative importance. It is an approach for multi-criteria decision, based on the outranking relation. Its evaluation considers the concepts of concordance and discordance. For each alternative (projects, in this work), these two indices are calculated, considering all criteria [24, 33].

Concordance index $C(a, b)$ identifies how much alternative $a$ is, at least, as good as alternative $b$. On the other hand, a measure of how much strictly preferable alternative $b$ is in comparison to alternative $a$ is done in discordance index $D(a, b)$ [33].

Next, some sets and quantities needed to calculate the two indices are presented. [33].

$$I^+ (a, b) = \left\{ C_i^+ \mid g_i(a) > g_i(b) \right\} \tag{1}$$

$$I^= (a, b) = \left\{ C_i^= \mid g_i(a) = g_i(b) \right\} \tag{2}$$

$$I^- (a, b) = \left\{ C_i^- \mid g_i(a) < g_i(b) \right\} \tag{3}$$

$$W^+ (a, b) = \sum_{j \in I^+(a,b)} w_i \tag{4}$$

$$W^= (a, b) = \sum_{j \in I^=(a,b)} w_i \tag{5}$$

$$W^- (a, b) = \sum_{j \in I^-(a,b)} w_i \tag{6}$$

where:

$i$ represents the $i^{th}$ selection criterion ($i = 1, ..., n$);

$g_i(j)$ indicates the preference value of the $i^{th}$ selection criterion for alternative $j$ ($j = 1, ..., m$);

$w_i$ is the weight of the $i^{th}$ selection criterion;

$\sum_{i=1}^{n} w_i = 1$;

$C_i^+ = 1$ if, alternative $a$ is strictly preferable in comparison to alternative $b$, otherwise $C_i^+ = 0$, for the $i^{th}$ selection criterion;

$C_i^= = 1$ if, alternative $a$ has the same preference in comparison to alternative $b$, otherwise $C_i^= = 0$, for the $i^{th}$ selection criterion;

$C_i^- = 1$ if, alternative $b$ is strictly preferable in comparison to alternative $a$, otherwise $C_i^- = 0$, for the $i^{th}$ selection criterion.

The concordance index $C(a, b)$ of alternative $a$ with respect to alternative $b$ is [33]:

$$C(a, b) = \frac{W^+(a, b) + W^=(a, b)}{W^+(a, b) + W^=(a, b) + W^-(a, b)} \tag{7}$$

The discordance index $D(a, b)$ of alternative $a$ with respect to alternative $b$ is [33]:

$$D(a, b) = \frac{max_{i \in I^-(a,b)} |g_i(a) - g_i(b)|}{max_{i \in I}(g_i^* - g_i^{**})} \tag{8}$$

where:

$g_i^*$ is the highest preference value for the $i^{th}$ selection criterion;

$g_i^{**}$ is the lowest preference value for the $i^{th}$ selection criterion.

For ranking the alternatives, decision-maker must compare the lists of concordance index, in descending order, discordance index, in ascending order.

Alternatively, a final ELECTRE II index could be calculated by:

$$e = C(a, b) - D(a, b) \tag{9}$$

The ranking is constructed ordering $e$ in descending order, from the highest (best option) to the lowest (worst option) values.

## 2.2 Preference Ranking Organization Method for Enrichment Evaluation II (PROMETHEE II)

PROMETHEE II is one of the existing MCDM (Multi-Criteria Decision Methods), used in decisions which there are conflicts between criteria, within a finite set of alternatives. It allows a ranking with the preferred options for each criterion [17]. Within the same criteria, the alternatives must be compared using the preference function $P$.

Let $f$ be a criterion with real value that needs to be maximized and $f(a)$ the evaluation of the alternative to this criterion. It must be compared to another alternative $b$ to obtain one that is preferred. Thus, we arrive at a preference function $P$. For each generalized criterion $f$, we have a corresponding preference function $P$, so that [6]:

$$P(a, b) = f(a) - f(b) \tag{10}$$

The function $H(d)$ related to the function $P$ can be considered [6, 17]:

$$H(d) = P(a, b) \text{ for } d > 0 \tag{11}$$

$$H(d) = 0 \text{ for } d \leqslant 0 \tag{12}$$

where:
$d = f(a) - f(b)$

Six different preference functions can be applied, according to how decision-maker intends to compare alternatives in a given criterion [6]. These functions are detailed in the sequence.

### Usual Criterion
In this case, there is only indifference between $a$ and $b$ if both alternatives are equal. In the event of a difference, the decision-maker has a strict preference for the largest.

$$H(d) = 0 \text{ for } d = 0 \tag{13}$$

$$H(d) = 1 \text{ for } d \neq 0 \tag{14}$$

### Almost Criterion
Considering a function $H(d)$ directly related to the preference function $P$, we have [6]:

$$H(d) = 0 \text{ for } d \leqslant q \tag{15}$$

$$H(d) = 1 \text{ for } d > q \tag{16}$$

where:
$q$ is the indifference threshold of the actions.

### Criterion with Linear Preference
The $p$ value is the one above which $d$ indicates a strict preference. If $d$ is less than $p$, preference of the decision-maker increases linearly with $d$ [6].

$$H(d) = 0 \text{ for } d \leqslant 0 \tag{17}$$

$$H(d) = \frac{d}{p} \text{ for } 0 < d \leqslant p \tag{18}$$

$$H(d) = 1 \text{ for } d > q \tag{19}$$

### Level Criterion
Decision-maker must define an indifference threshold $q$ and a preference threshold $p$ [6].

$$H(d) = 0 \text{ for } d \leqslant q \tag{20}$$

$$H(d) = \frac{1}{2} \text{ for } q < d \leqslant p \tag{21}$$

$$H(d) = 1 \text{ for } d > q \tag{22}$$

### Criterion with Linear Preference and Indifference Area
Bearing in mind that the indifference threshold $q$ is the highest value of $d$, below which there is indifference, and the preference threshold $p$ is the lowest value of $d$ above which there is strict preference. If $d$ is less than $p$ greater than $q$, the preference of the decision-maker increases linearly with $d$ [6].

$$H(d) = 0 \text{ for } d \leqslant q \tag{23}$$

$$H(d) = \frac{d - q}{p - q} \text{ for } q < d \leqslant p \tag{24}$$

$$H(d) = 1 \text{ for } d > p \tag{25}$$

The decision-maker linearly increases indifference to a specific preference [33].

**Gaussian Criterion**

In this function, there are no discontinuities and the stability of results is guaranteed [6].

$$H(d) = 1 - e^{-d^2/2\sigma^2} \tag{26}$$

where:

$\sigma^2$ is the variance of a normal distribution.

The choice of appropriate preference function depends on the decision to be made. To find the ranking of options, it is necessary to follow five steps [6, 17, 33]:

**Step 1** - Calculate the deviations between a pair of criteria:

$$d_i(a, b) = g_i(a) - g_i(b) \tag{27}$$

where:

$g_i(a)$ and $g_i(b)$ are the evaluations of alternatives $a$ e $b$, respectively, both belonging to set $A$ in criterion $i$ ($i = 1, ..., n$);

$A$ is the set of alternatives to be compared among criteria.

**Step 2** - Apply the preference function $P$:

$$P_i(a, b) = F_i[g_i(a, b)] \tag{28}$$

**Step 3** - Calculate global preference index ($\pi$):

$$\forall a, b \in A \rightarrow \pi(a, b) = \sum_{i=1}^{n} w_i P_i(a, b) \tag{29}$$

where:

$w_i$ is the weight of criterion $i$;

$\sum_{i=1}^{n} w_i = 1$.

**Step 4** - Calculate the overshoot flows for each alternative ($\phi$), positive and negative:

$$\phi^+(a) = \sum_{x \in A}^{m} \pi(a, x) \tag{30}$$

$$\phi^-(a) = \sum_{x \in A}^{m} \pi(x, a) \tag{31}$$

where:

$x \in A | x \neq a$

$m$ is the total of alternatives.

**Step 5** - Calculate the net overshoot flows for each alternative:

$$\phi(a) = \phi^+(a) - \phi^-(a) \tag{32}$$

The ranking is built ordering $\phi$ index, of each alternative, from the highest (the best option) to the lowest (the worst option).

## 2.3   Vlsekriterijumska Optimizacija I Kompromismo Resenje (VIKOR)

VIKOR (in English, Multicriteria Optimization and Compromise Solution) is a multi-criteria decision method that can also be used in problems that present conflicting criteria due to its importance, with goal to find the best and worst values, through rankings, so that the alternative chosen is the closest to ideal [17, 33].

Classification using VIKOR multi-criteria decision method follows these steps [3]:

**Step 1** - Determine the best and the worst alternatives ($f_i^*$ e $f_i^-$, respectively), for each criterion $i$ ($i = 1, 2, ..., n$).

**Step 2** - Calculate, for each alternative $j$ ($j = 1, 2, .., m$), $S$ and $R$ values:

$$S_j = \sum_{i=1}^{n} w_i \frac{f_i^* - f_{ij}}{f_i^* - f_i^-} \tag{33}$$

$$R_j = \max[w_i \frac{f_i^* - f_{ij}}{f_i^* - f_i^-}] \tag{34}$$

where:
$w_i$ is the weight of criterion $i$;
$\sum_{i=1}^{n} w_i = 1$.

**Step 3** - Calculate:

$$Q_j = [v \frac{(S_j - S^*)}{S^- - S^*}] + [(1 - v)\frac{(R_j - R^*)}{R^- - R^*}] \tag{35}$$

where:
$S^* = \min_j S_j$
$S^- = \max_j S_j$
$R^* = \min_j R_j$
$R^- = \max_j R_j$
$v$ = strategy, being able to adopt 0.5 if there is no preference between $S$ e $R$.

**Step 4** - Sort alternatives using values of $S$, $R$ e $Q$ in descending order.

**Step 5** - The possible solution will be the alternative $j_1$ classified with the lowest $Q$, if the following conditions are met:

- Acceptable Advantage

$$Q(j_2) - Q(j_1) \geqslant DQ \tag{36}$$

where:
$j_1$ and $j_2$ are alternatives that are in the first and second position, respectively, in $Q$ ranking list;
$DQ = 1/(m-1)$;

- Acceptable Stability in Decision-Making

The alternative $a$ should be rated from the best $S$ or $R$. In decision-making process, this solution is stable by voting on the majority rule ($v > 0.5$), by consensus ($v \approx 0.5$) or by veto ($v < 0.5$).

If Acceptable Advantage and Acceptable Stability conditions are not met, a set of compromise solutions is proposed, according to:

- $j_1$ and $j_2$ if only the condition of Acceptable Stability has not been met;
- $j_1, j_2,..., j_M$ if the Acceptable Advantage condition has not been met. $j_M$ is determined by: $Q(j_M) - Q(j_1) < DQ$, with $M$ being value of the maximum position of the alternative in the classification.

The classification result is a list of alternatives balancing commitment and the compromise solution [23].
VIKOR multi-criteria decision method assists in situations which the decision-maker is unable to express his preferences at the project beginning. The compromise solution found is acceptable for decision, as it provides minimum $S$ and $R$.

### 2.4  Technique for Order Preference by Similarity to Ideal Solution (TOPSIS)

TOPSIS is a multi-criteria method focused on measuring the performance of possible solutions, comparing them. The best alternative is the one closest to ideal solution and the furthest from the non-ideal solution. Each criterion must have a weight, according to what the decision-maker considers important [17, 33].
The process of using the TOPSIS method consists of performing the following steps [17, 33].

**Step 1** - Calculate the normalized $r_{ij}$ in the decision matrix:

$$r_{ij} = \frac{f_{ij}}{\sqrt{\sum_{j=1}^{m} f_{ij}^2}} \tag{37}$$

where:
$j$ is the alternative ($j = 1, ..., m$)
$i$ is the selection criteria ($i = 1, ..., n$)

**Step 2** - Calculate the weighted $\varsigma^{ij}$ for each $r_{ij}$ in normalized decision matrix:

$$\varsigma^{ij} = w_i r_{ij} \tag{38}$$

where:

$w_i$ is the weight of $i^{th}$ criterion;

$\sum_{i=1}^{n} w_i = 1.$

**Step 3** - Determine the ideal and negative solution:

$$\varsigma^* = (\max_j)\varsigma_{ij}|j \in J', (\min_j)\varsigma_{ij}|j \in J'' \tag{39}$$

$$\varsigma^- = (\min_j)\varsigma_{ij}|j \in J', (\max_j)\varsigma_{ij}|j \in J'' \tag{40}$$

where:

$J'$ associated with positive impact criteria;

$J''$ associated with negative impact criteria.

**Step 4** - Use Euclidean distance to calculate separation measures ($D_j^*$ and $D_j^-$ separations from the ideal and negative solutions, respectively):

$$D_j^* = \sqrt{\sum_{i=1}^{n}(\varsigma_{ij} - \varsigma_i^*)^2} \tag{41}$$

$$D_j^- = \sqrt{\sum_{i=1}^{n}(\varsigma_{ij} - \varsigma_i^-)^2} \tag{42}$$

**Step 5** - Calculate the relative proximity of the ideal solution:

$$C_j^* = \frac{D_j^-}{D_j^* - D_j^-} \tag{43}$$

**Step 6** - Catalog the order of preference, considering the increasing values of $C_j^*$. Through this ranking of preferences, it is possible to define the optimal alternative [23].

## 3   Uncertainty

Projects selection may consider the uncertainty [8]. When applying MCDM, it is necessary that decision-maker evaluates and estimates the values of criteria for each project. Uncertainty is an inherent effect of estimating and forecasting [4, 16].

When estimating a value for a parameter, the responsible person chooses the one that best represents their assessment. However, there may not be precision in this choice [11]. Or even, if several people estimate values for the same parameter, their estimates may differ from each other.

Take as an example three experts estimating values for the same selection criteria, for a given alternative. Most likely, instead of a single value, what one will get is a range of values. This illustrates the uncertainty in choosing the parameter values used in a selection, which can impact the final decision.

Uncertainty in project selection may change the decision, which must be worse in R&D development, due to its nature, with long planning and development period [9, 13, 16]. For minimizing this impact, instead of working with a single value, three-point estimation can be useful. It is done by estimating the values most likely, best-case and worst-case. The most likely value is the estimation of the parameter best understanding by evaluator (probably, this is the value estimated if a single point estimation is proceeded). Best-case and worst-case values must reflect the best and worst scenarios, respectively [14, 26].

These three values allow to construct a triangular probability distribution, as suggested by several works in literature [15, 18, 26, 32], and presented in Fig. 1:

- parameter $a$ is equal to worst-case estimation;
- parameter $b$ is equal to best-case estimation;
- parameter $c$ is equal to most likely estimation.

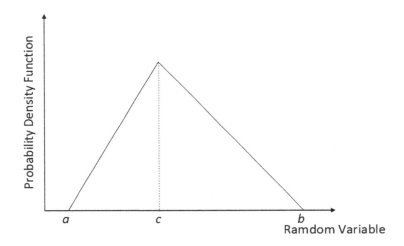

**Fig. 1.** Triangular probability distribution based on three point estimation.

Defined triangular distribution of each estimated value, a Monte Carlo simulation can be used to adequately address uncertainty in the decision process [15, 16, 30]. It executes multiple simulation rounds and the final decision can be based on the mean of results.

## 4   Proposed Method

This paper proposes a method for assisting decision-makers in project selection. Its ultimate goal is to define a ranked list, ranging from the best to the worst, to aid in the decision-making process. It's based on MCD Methods and takes into account the uncertainty that comes with estimating criteria values.

Three values are estimated for each criterion for each project. As explained in 3, triangular probability distributions are created and Monte Carlo simulations are undertaken. Decision-makers must also set the weights of criteria.

Uncertainty in selection is handled using Monte Carlo simulation, in which every round ($z$ rounds), project parameters are randomly chosen based on the triangular distributions. For avoiding distortion due to different value ranges, they are normalized. Then ELECTRE II, PROMETHEE II, VIKOR and TOPSIS evaluations are executed. At the end of rounds, there are $z$ sets of $e$, $\phi$, $\pi$ and $Q$ indeces calculated, defining final values of each one by calculating the means.

These four methods may indicate different rankings among them, once they differ in calculation procedures. This work proposes a solution with a fifth index (called Integrated Index, normalized between 0 and 1) by the mean of $e$, $\phi$, $\pi$ and $Q$ normalized indeces. The normalization of values before calculating the Integrated Index aims to ensure that one method does not stand out from the other in the final ranking. It allows a more robust decision, given that its result considers the four methods together. Figure 2 presents proposed method flow.

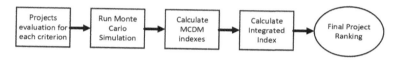

**Fig. 2.** Proposed method flow.

The algorithm for ranking projects, with a stochastic approach to address uncertainty in the definition of parameters and using Monte Carlo simulation is the following ($z$ is the number of Monte Carlo rounds):

```
Begin

Define z

Read worst-case_project_estimation

Read most_likely_project_estimation

Read best-case_project_estimation

Read criteria_weigths

Repeat z times

{

Define ramdomly project_parameters
```

```
Normalize values

Calculate e_index / phi_index / pi_index / Q_index

Store e_index / phi_index / pi_index / Q_index

}

Calculate mean of z sets of e_index values
Calculate mean of z sets of phi_index values
Calculate mean of z sets of pi_index values
Calculate mean of z sets of Q_index values

Normalize e_index / phi_index / pi_index / Q_index
Calculate Integrated_index

Define final ranking

End
```

## 5    Numerical Example

The strategy proposed in Sect. 4 was used in a set of ten real hardware and software development projects candidates as an example, from R&D department of an information technology (IT) equipment provider company. Specialists on IT projects and market from the company estimated values for four criteria:

**C1 - Competitiveness improvement (weight - 0,3):** the capacity of project for improving company competitiveness (from 1 - the lowest to 10 - the highest);

**C2 - Market Potential (weight - 0,2):** the capacity of project for improving market share or market insertion (from 1 - the lowest to 10 - the highest);

**C3 - Degree of Innovation (weight - 0,2):** how innovative the project is (from 1 - the lowest to 10 - the highest).

**C4 - Return/risk Rate (weight - 0,3):** an evaluation of the ratio between the estimated return and the associated risk (from 1 - the lowest to 10 - the highest);

The objective was to select three projects for execution. They should be chosen in order to be more aligned with the company's strategies and met the defined criteria in the best way. Preparing a ranking (based on the criteria and using ELECTRE II, PROMETHEE II, VIKOR and TOPSIS methods), the three best ranked ones should be executed.

The first step was the definition, by the specialists consulted, of each criterion values for the projects. One by one, they defined the most likely (ML), worst-case (WC) and best-case (BC) values, of the criteria defined for selection, presented in Table 1. Most likely values are generally used when single estimation is applied in selection. Worst-case and best-case values allow applying the selection considering uncertainty with triangular probability distribution.

**Table 1.** Projects characteristics.

| Project | Criteria | | | | | | | | | | | |
|---|---|---|---|---|---|---|---|---|---|---|---|---|
| | C1 | | | C2 | | | C3 | | | C4 | | |
| | WC | ML | BC | WC | ML | BC | WC | ML | BC | WC | ML | BC |
| A | 1 | 1 | 3 | 9 | 10 | 10 | 3 | 3 | 5 | 6,9 | 7,1 | 9,0 |
| B | 7 | 9 | 9 | 3 | 5 | 5 | 2 | 4 | 4 | 6,7 | 8,2 | 8,3 |
| C | 8 | 10 | 10 | 6 | 8 | 10 | 2 | 4 | 6 | 2,2 | 2,7 | 4,4 |
| D | 4 | 5 | 5 | 8 | 9 | 10 | 4 | 5 | 5 | 8,4 | 9,0 | 10,0 |
| E | 2 | 3 | 5 | 1 | 1 | 3 | 8 | 9 | 10 | 1,0 | 1,2 | 1,8 |
| F | 6 | 8 | 8 | 5 | 7 | 8 | 7 | 9 | 9 | 0,3 | 1,0 | 1,1 |
| G | 4 | 4 | 5 | 8 | 8 | 9 | 1 | 1 | 2 | 3,6 | 4,5 | 6,2 |
| H | 7 | 9 | 10 | 1 | 1 | 2 | 4 | 6 | 6 | 3,7 | 5,6 | 5,7 |
| I | 1 | 2 | 4 | 1 | 3 | 5 | 4 | 6 | 8 | 6,5 | 6,7 | 8,0 |
| J | 7 | 9 | 9 | 3 | 5 | 5 | 1 | 3 | 3 | 2,2 | 2,2 | 2,4 |

Before simulation using the proposed method, one was made using the single estimation (deterministic evaluation with values presented in columns ML in Table 1) as a reference for comparing the results. In this case, selected projects would be B, D, and H. They were the best ranked by Integrated Index, as can be seen in graphic of Fig. 3, presenting the best option on the left, and the worst on the right. This graphic also presents B and D Integrated Index values are too close, with B a little better.

Applying the selection with uncertainty, as presented in Sect. 4 (Monte Carlo simulation with 10,000 rounds), the ranking changed, as shown in Table 2 and in Fig. 4, indicating projects D, B and C for execution. A different set, compared to the deterministic evaluation.

In all applications of PROMETHEE method in this work, preference function was Criterion with linear preference and indifference area.

## 5.1 Results Analysis

Comparing the results among the four methods (ELECTRE II, PROMETHEE II, VIKOR, and TOPSIS), presented in Table 2, some differences are identified. Three of them indicated the same best project option (D), but VIKOR method indicated B as the best option. But all of them indicated the same set for selection. There are also some inversions in fifth, sixth, seventh, eighth, and ninth positions among the methods As they have the same purpose of selection, it indicates the importance of considering all four together in a single Integrated Index, as proposed in this work.

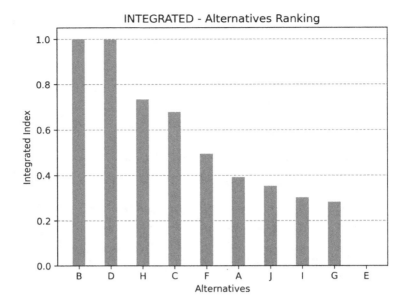

**Fig. 3.** Projects ranking - integrated index - without uncertainty.

**Table 2.** Results.

| Projects | PROMETHEE II | | TOPSIS | | VIKOR | | ELECTRE II | | Integrated | |
|---|---|---|---|---|---|---|---|---|---|---|
| | $\phi$ | Order | $\pi$ | Order | Q | Order | e | Order | Index | Order |
| A | 0,06 | 6 | 0,50 | 5 | 0,72 | 8 | 0,06 | 5 | 0,45 | 5 |
| B | 1,31 | 2 | 0,63 | 2 | 0,07 | 1 | 2,68 | 2 | 0,91 | 2 |
| C | 1,05 | 3 | 0,56 | 3 | 0,33 | 3 | 2,30 | 3 | 0,75 | 3 |
| D | 1,84 | 1 | 0,64 | 1 | 0,11 | 2 | 4,46 | 1 | 1,00 | 1 |
| E | -2,11 | 10 | 0,36 | 10 | 0,94 | 10 | -4,08 | 10 | 0,00 | 10 |
| F | 0,27 | 4 | 0,49 | 6 | 0,70 | 7 | -0,44 | 6 | 0,45 | 6 |
| G | -0,89 | 8 | 0,42 | 9 | 0,53 | 5 | -2,21 | 9 | 0,31 | 7 |
| H | 0,23 | 5 | 0,54 | 4 | 0,37 | 4 | 0,92 | 4 | 0,63 | 4 |
| I | -0,87 | 7 | 0,46 | 7 | 0,75 | 9 | -1,53 | 7 | 0,29 | 8 |
| J | -0,90 | 9 | 0,44 | 8 | 0,67 | 6 | -2,15 | 8 | 0,28 | 9 |

Comparing the results without and with uncertainty, we observe more additional differences. The sets for selection changed from deterministic evaluation (B, D, and H) to stochastic one (D, B, and C). There is also change in fourth position in ranking: C (deterministic) or H (stochastic). When analyzing method by method, results indicate other differences among projects ranking position. The methods TOPSIS and ELECTRE II coincided in all ranking positions, but there were divergences with PROMETHEE II and VIKOR methods. The difference in the best ranked option is also highlighted, depend-

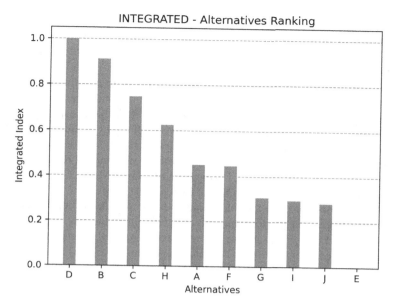

**Fig. 4.** Projects ranking - integrated index - with uncertainty.

ing on whether or not uncertainty is considered (from D to B, respectively). These changes showed the importance of considering uncertainty in this kind of selection.

Finally, the proposed method delivers as outcome the recommended set of projects (portfolio) to be executed, being the one obtained from the stochastic scenario: D, B and C. It considers the uncertainty and all four methods together in the selection.

## 6  Conclusions

Companies face a challenge when they need to select their projects for execution. Usually, there are not enough resources available to run all the candidates. In view of this, decision-makers need to choose the portfolio of projects to be executed, considering factors such as strategic alignment with the company's objectives, estimated return and associated risk.

For decision-making, generally several criteria must be evaluated simultaneously, which makes the decision more complex. When decision is based on only one criterion, one just compares the evaluation for it and decide for the best option(s). But, with more than one criterion, the analysis must consider all of them at the same time. Additionally, some of them can also be in conflict with each other, increasing difficulty in decision process.

In such situations, multi-criteria decision methods (MCDM) help decision-makers in selection. Methods as *Elimination Et Choix Traduisant la Réalité* II (ELECTRE II), Preference Ranking Organization Method for Enrichment Evaluation II (PROMETHEE II), *Vlsekriterijumska Optimizacija I Kompromismo Resenje* (VIKOR), and Technique for Order Preference by Similarity to Ideal Solution (TOPSIS) allow the simultaneous comparison among different and conflicting criteria, allowing the elaboration of a ranking in order of preference for execution. Based on this ranking order, one can select projects to be included in company's portfolio for execution.

As they base their decision on estimated values, there is no certainty about the exact values of the criteria evaluated. In this scenario, uncertainty must be considered in the process, since estimates inherently bring uncertainties with them. Especially in R&D projects, whose characteristics include long periods of planning and execution.

This study suggests a method for incorporating uncertainty into decision and assisting decision-makers. Instead of utilizing a single value to estimate, three-point are used: most likely, worst-case, and best-case. On the basis of them, a Monte Carlo simulation is run, with the parameters randomly defined in each round (in this case, applying triangular probability distribution). Because these methods can produce different rankings for the same set of projects, the proposal recommends combining the four MCDM (creating an Integrated Index) for a final ranking, allowing for a more robust selection outcome.

The numerical example used was the selection of three projects, among ten candidates for execution in an IT equipment provider company. Without taking into account the uncertainty, the result chooses projects B, D, and H. However, the indication changed when the uncertainty was considered, proposing D, B, and C, this being the set (portfolio) to be executed in the company. Another point observed was that, considering the methods individually, there were divergence in the list of selected projects and in the ranking order. These results reinforce the importance of considering uncertainty in the decision-making process and integrated method, presenting a single final ranking.

For future works, some issues must be considered in selection:

- Apply a fuzzy approach to address uncertainty rather than Monte Carlo simulation;
- Constraints as developers and equipment available;
- Time needed for each project execution.

# References

1. Abbassi, M., Ashrafi, M., Tashnizi, E.S.: Selecting balanced portfolios of R&D projects with interdependencies: a cross-entropy based methodology. Technovation **34**(1), 54–63 (2014)
2. Agapito, A.O., Vianna, M.F.D., Moratori, P.B., Vianna, D.S., Meza, E.B.M., Matias, I.O.: Using multicriteria analysis and fuzzy logic for project portfolio management. Braz. J. Oper. Prod. Manage. **16**(2), 347–357 (2019)
3. Archer, N.P., Ghasemzadeh, F.: An integrated framework for project portfolio selection. Int. J. Project Manage. **17**(4), 207–216 (1999)
4. Bohle, F., Heidling, E., Schoper, Y.: A new orientation to deal with uncertainty in projects. Int. J. Project Manage. **34**(7), 1384–1392 (2015)
5. Brans, J.P., Vincke, P.: A preference ranking organisation method: (the PROMETHEE method for multiple criteria decision-making). Manage. Sci. **31**(6), 647–656 (1985)
6. Brans, J.P., Vincke, P., Mareschal, B.: How to select and how to rank projects: the PROMETHEE method. Eur. J. Oper. Res. **24**(2), 228–238 (1986)
7. Dutra, C.C., Ribeiro, J.L.D., de Carvalho, M.M.: An economic-probabilistic model for project selection and prioritization. Int. J. Project Manage. **32**(6), 1042–1055 (2014)

8. Eilat, H., Golany, B., Shtub, A.: Constructing and evaluating balanced portfolios of R&D projects with interactions: a DEA based methodology. Eur. J. Oper. Res. **172**(3), 1018–1039 (2006)
9. Hassanzadeh, F., Modarres, M., Nemati, H.R., Amoako-Gyampah, K.: A robust R&D project portfolio optimization model for pharmaceutical contract research organizations. Int. J. Prod. Econ. **158**, 18–27 (2014)
10. Hwang, C.L., Yoon, K.: Multiple Attribute Decision Making: Methods and Applications. Springer-Verlag, Nova York, EUA (1981). https://doi.org/10.1007/978-3-642-48318-9
11. Jaafari, A.: Management of risks, uncertainties and opportunities on projects: time for a fundamental shift. Int. J. Project Manage. **19**(2), 89–101 (2001)
12. Jo, S.H., Lee, E.B., Pyo, K.Y.: Integrating a procurement management process into critical chain project management (CCPM): a case-study on oil and gas projects, the piping process. Sustainability **10**(6), 1817 (2018)
13. Lee, S., Cho, Y., Ko, M.: Robust optimization model for R&D project selection under uncertainty in the automobile industry. Sustainability **12**(23), 773–788 (2020)
14. Marcondes, G.A.B.: Multicriteria decision method for project ranking considering uncertainty. In: International Conference on Operations Research and Enterprise Systems - ICORES (2021)
15. Marcondes, G.A.B., Leme, R.C., de Carvalho, M.M.: Framework for integrated project portfolio selection and adjustment. IEEE Trans. Eng. Manage. **66**(4), 677–688 (2019)
16. Marcondes, G.A.B., Leme, R.C., Leme, M.S., da Silva, C.E.S.: Using mean-Gini and stochastic dominance to choose project portfolios with parameter uncertainty. Eng. Econ. **62**(1), 33–53 (2017)
17. Martins, D.T., Marcondes, G.A.B.: Project portfolio selection using multi-criteria decision methods. In: IEEE International Conference on Technology and Entrepreneurship - ICTE (2020)
18. Mathews, S.: Valuing risky projects with real options. Res. Technol. Manag. **52**(5), 32–41 (2009)
19. Mavrotas, G., Makryvelios, E.: Combining multiple criteria analysis, mathematical programming and Monte Carlo simulation to tackle uncertainty in research and development project portfolio selection: a case study from Greece. Eur. J. Oper. Res. **291**(2), 794–806 (2021)
20. Miranda, M.M., Raymond, J., Dezayes, C.: Uncertainty and risk evaluation of deep geothermal energy source for heat production and electricity generation in remote northern regions. Energies **13**(16), 4221 (2020)
21. Mousavi, S.A., Seiti, H., Hafezalkotob, A., Asian, S., Mobarra, R.: Application of risk-based fuzzy decision support systems in new product development: an R-VIKOR approach. Appl. Soft Comput. **107**, 107456 (2021)
22. Opricovic, S.: Multi-criteria optimization of civil engineering systems (in Serbian, Visekriterijumska optimizacija sistema u gradjevinarstvu). Ph.D. thesis, Faculty of Civil Engineering, Belgrade (2012)
23. Opricovic, S., Tzeng, G.H.: Compromise solution by MCDM methods: a comparative analysis of VIKOR and TOPSIS. Eur. J. Oper. Res. **156**(2), 445–455 (2004)
24. Opricovic, S., Tzeng, G.H.: Extended VIKOR method in comparison with outranking methods. Eur. J. Oper. Res. **178**(2), 514–529 (2006)
25. Perez, F., Gomez, T.: Multiobjective project portfolio selection with fuzzy constraints. Ann. Oper. Res. **245**(1), 7–29 (2014). https://doi.org/10.1007/s10479-014-1556-z
26. PMI: A Guide to the Project Management Body of Knowledge. Project Management Institute, Atlanta, EUA, 6 edn. (2017)
27. Riley, D., Schaafsma, M., Marin-Moreno, H., Minshull, T.A.: A social, environmental and economic evaluation protocol for potential gas hydrate exploitation projects. Appl. Energy **263**, 114651 (2020)

28. Roy, B., Bertier, P.: La Methode ELECTRE II: Une application au media-planning. Oper. Res. 291–302 (1973)

29. Sadi-Nezhad, S.: A state-of-art survey on project selection using MCDM techniques. J. Project Manage. 2(1), 1–10 (2017)

30. Shakhsi-Niaei, M., Torabi, S., Iranmanesh, S.: A comprehensive framework for project selection problem under uncertainty and real-world constraints. Comput. Ind. Eng. 61(1), 226–237 (2011)

31. Song, Y., Li, X., Li, Y., Hong, X.: Assessing the risk of an investment project using an improved TOPSIS method. Appl. Econ. Lett. 27(16), 1334–1339 (2020)

32. Stein, W.E., Keblis, M.F.: A new method to simulate the triangular distribution. Math. Comput. Model. 49(5–6), 1143–1147 (2009)

33. Tzeng, G.H., Huang, J.J.: Multiple Attribute Decision Making: Methods and Applications. Chapman and Hall/CRC (2011)

34. Urli, B., Terrien, F.: Project portfolio selection model, a realistic approach. Int. Trans. Oper. Res. 17(6), 809–826 (2010)

35. Wallenius, J., Dyer, J.S., Fishburn, P.C., Steuer, R.E., Zionts, S., Deb, K.: Multiple criteria decision making, multiattribute utility theory: recent accomplishments and what lies ahead. Manage. Sci. 54(7), 1336–1349 (2008)

36. Wu, Y., Zhang, B., Zhang, C.W.T., Liu, F.: Optimal site selection for parabolic trough concentrating solar power plant using extended PROMETHEE method: a case in China. Renew. Energy 143, 1910–1927 (2019)

37. Yang, I.T.: Impact of budget uncertainty on project time-cost tradeoff. IEEE Trans. Eng. Manag. 52(2), 167–174 (2005)

38. Yu, S., Zhang, S., Agbemabiese, L., Zhang, F.: Multi-stage goal programming models for production optimization in the middle and later periods of oilfield development. Ann. Oper. Res. 255(1–2), 421–437 (2017)

# Finite-Horizon and Infinite-Horizon Markov Decision Processes with Trapezoidal Fuzzy Discounted Rewards

Karla Carrero-Vera[1]([⊠])(iD), Hugo Cruz-Suárez[1]([⊠])(iD), and Raúl Montes-de-Oca[2]([⊠])(iD)

[1] Facultad de Ciencias Físico Matemáticas, Benemérita Universidad Autónoma de Puebla, Av. San Claudio y Río Verde, Col. San Manuel, CU, 72570 Puebla, Puebla, Mexico
karla.carrero@alumno.buap.mx, hcs@fcfm.buap.mx
[2] Departamento de Matemáticas, Universidad Autónoma Metropolitana-Iztapalapa, Av. San Rafael Atlixco 186, Col. Vicentina, 09340 Ciudad de México, Mexico
momr@xanum.uam.mx

**Abstract.** Discrete-time discounted Markov decision processes (MDPs, in singular MDP) with finite state spaces, compact action sets and trapezoidal fuzzy reward functions are presented in this article. For such a kind of MDPs, both the finite and the infinite horizons cases are studied. The corresponding optimal control problems are established with respect to the partial order on the $\alpha$-cuts of fuzzy numbers, named the fuzzy max order. The fuzzy optimal solution is related to a suitable discounted MDP with a nonfuzzy reward. And in the article, different applications of the theory developed are provided: a finite-horizon model of an inventory system in which an algorithm to calculate the optimal solution is given, and, additionally for the infinite-horizon case, an MDP and a competitive MDP (also known as a stochastic game) are supplied in an economic and financial context.

**Keywords:** Discounted Markov decision process · Optimal policy · Fuzzy set · Trapezoidal fuzzy number · Fuzzy reward

## 1 Introduction

The article concerns discrete-time Markov decision processes (MDPs, in singular MDP) [14,21]. Concretely, MDPs with a finite state space, compact action sets and with a discounted reward as the objective function are dealt with, and both the finite-horizon and the infinite-horizon problems are considered. Moreover, it is important to note that as appears in the specialized literature (see, for instance, [17]), the main motivation to analyze such a kind of MDPs is predominantly economic.

This work is also related to the well-known fuzzy set theory which has been established in 1965 by L.A. Zadeh [28], and as it can be seen in the following references, this topic has been widely developed both at a theoretical and applied level: [5,7–9,13,15] and [18].

Now, for an MDP with the characteristics mentioned in the first paragraph, which will be named the standard MDP, the relation to the fuzzy theory consists in changing

© Springer Nature Switzerland AG 2022
G. H. Parlier et al. (Eds.): ICORES 2020/2021, CCIS 1623, pp. 171–192, 2022.
https://doi.org/10.1007/978-3-031-10725-2_9

in the decision model of this MDP the corresponding reward function for a trapezoidal fuzzy (see [1,6]) function which models the possible imprecision or inexactness that the real data impose on the standard reward, and the idea is to study the new optimal control problem with this fuzzy reward. In summary, firstly, a standard discounted MDP [14,21] is considered; secondly, in the corresponding decision model of this standard MDP, the reward $R$ is substituted by a suitable trapezoidal fuzzy function $\tilde{R}$, where $\tilde{R}$ is adequately considered as a function of $R$; then, with this, an optimal control problem with a trapezoidal fuzzy discounted objective function is induced, where, taking into account that the set of strategies for both MDPs, the standard and the fuzzy one coincides, the optimum is taken with respect to the max order of the fuzzy numbers (see [11]).

The main results obtained are that the optimal strategy for both the standard MDP and the fuzzy one are the same, and that the fuzzy optimal value function is of trapezoidal type.

To exemplify the theory developed a finite-horizon inventory control system [19] with a suitable algorithm to obtain the optimal policy is given, and, for the infinite-horizon case, a portfolio choice problem [17,27] and its extension to a two-person stochastic game [10,24] are also provided.

In the fuzzy theory, the class of trapezoidal functions is very relevant (see [1,6]), and in particular, this class contains the class of triangular fuzzy functions. Therefore, the present work is a substantial extension of [4] in which a study of discounted MDPs with triangular fuzzy rewards is presented.

Additional antecedents for this paper are [16] and [23]. In these papers infinite-horizon discounted MDPs on finite spaces and with fuzzy rewards are studied.

The article is organized as follows. Section 2 gives some basics on fuzzy theory. Section 3 provides the results on standard discounted MDPs, while Sect. 4 considers MDPs with fuzzy rewards. Section 5 presents the examples and, finally, a concluding remark is given in Sect. 6.

## 2    Preliminaries: Fuzzy Sets

This section presents some definitions and basic results about the fuzzy set theory [15, 25]. These facts will be used in Sect. 4.

### 2.1    Arithmetic

The fuzzy set theory was proposed by Zadeh in 1965 [28], an interesting feature of using a fuzzy approach is that it allows the use of linguistic variables such as: "low", "very", "high", "advisable", "highly risky", etc. To introduce the concept of fuzzy number, firstly, the crisp set definition will be presented.

**Definition 1.** *A crisp set is defined as a collection of elements or objects, which can be finite, countable, or uncountable.*

**Definition 2.** *Let $\Theta$ be a non-empty set. Then a fuzzy set $A$ on $\Theta$ is defined in terms of the* membership function $\mu : \Theta \longrightarrow [0,1]$. *Consequently, a fuzzy set $A$ is defined as a set of ordered pairs: $\{(x, \mu(x)) : x \in \Theta\}$.*

*Remark 1.* In Definition 2, $\mu(x)$ represents the degree of compatibility or degree of truth to which the element $x$ verifies the characteristic property of a set $A \subset \Theta$.

Then, using the membership function, a fuzzy number can be defined as follows.

**Definition 3.** *A fuzzy number $A$ is a fuzzy set defined on the real numbers $\mathbb{R}$ characterized by means of a membership function $\mu$, $\mu : \mathbb{R} \longrightarrow [0,1]$,*

$$\mu(x) = \begin{cases} 0, & x \le a \\ l(x), & a < x \le b \\ 1, & b < x \le c \\ r(x), & c < x \le d \\ 0, & d < x, \end{cases} \tag{1}$$

*where $a, b, c$, and $d$ are real numbers, $l$ is a non-decreasing function and $r$ is a non-increasing function. The functions $l$ and $r$ are called the left and right sides of the fuzzy number $A$, respectively.*

In this article the authors will focus on the study of trapezoidal fuzzy numbers, which are presented in the following definition, as a consequence of Definition 3.

**Definition 4.** *A fuzzy number $A$ is called a* trapezoidal fuzzy number *if its membership function has the following form:*

$$\mu(x) = \begin{cases} 0, & x < \gamma_1 \\ \dfrac{x - \gamma_1}{\gamma_2 - \gamma_1}, & \gamma_1 \le x < \gamma_2 \\ 1, & \gamma_2 \le x \le \gamma_3 \\ \dfrac{\gamma_4 - x}{\gamma_4 - \gamma_3}, & \gamma_3 < x \le \gamma_4 \\ 0, & x > \gamma_4, \end{cases} \tag{2}$$

*i.e. making $l(x) = \dfrac{x - \gamma_1}{\gamma_2 - \gamma_1}$ and $r(x) = \dfrac{\gamma_4 - x}{\gamma_4 - \gamma_3}$ in (1), where $\gamma_1$, $\gamma_2$ and $\gamma_3$ are real numbers such that $\gamma_1 < \gamma_2 \le \gamma_3 < \gamma_4$. In the rest of this work, a trapezoidal fuzzy number is denoted by $\mu = (\gamma_1, \gamma_2, \gamma_3, \gamma_4)$.*

*Example 1.* Figure 1 illustrates a graphical representation of a trapezoidal fuzzy number $A = (1/2, 3, 5, 7)$.

**Definition 5.** *Let $A$ be a fuzzy number with a membership function $\mu$ and let $\alpha$ be a real number of the interval $[0, 1]$. Then the $\alpha$-cut of $A$, denoted by $\mu_\alpha$, is defined to be the set $\{x \in \mathbb{R} : \mu(x) \ge \alpha\}$ for $\alpha > 0$ and $\mu_0 := cl\{x \in \mathbb{R} : \mu(x) > 0\}$, where $cl$ denotes the closure of the set $\{x \in \mathbb{R} : \mu(x) > 0\}$.*

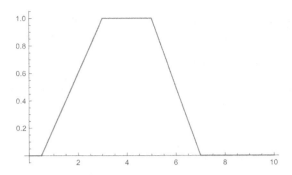

**Fig. 1.** A trapezoidal fuzzy number.

*Remark 2.* According to Definitions 3 and 5, for a fuzzy number $A$, its $\alpha$-cut set, $\mu_\alpha = [A^-(\alpha), A^+(\alpha)]$ is a closed interval, where $A^-(\alpha) = \inf\{x \in \mathbb{R} : \mu_A(x) \geq \alpha\}$ and $A^+(\alpha) = \sup\{x \in \mathbb{R} : \mu_A(x) \geq \alpha\}$.

*Example 2.* Consider the trapezoidal fuzzy number $\mu = (\gamma_1, \gamma_2, \gamma_3, \gamma_4)$. Then, from Definition 5, it is easy to prove that the $\alpha$-cut of $\mu$ is given by the closed interval:

$$\mu_\alpha = [\gamma_1 + \alpha(\gamma_2 - \gamma_1), \gamma_4 - \alpha(\gamma_4 - \gamma_3)],$$

for $0 \leq \alpha \leq 1$.

In the subsequent, it will be denoted by $\mathfrak{F}(\Theta)$ the set of all convex fuzzy sets whose membership functions are upper-semicontinuous, normal and have a compact support. In particular, the manuscript is focused on the cases when $\Theta$ is the crisp set of real numbers $\mathbb{R}$, in this case, $\mathfrak{F}(\mathbb{R})$ denotes the set of all fuzzy numbers. (See [15] for the proof that each element of $\mathfrak{F}(\mathbb{R})$ satisfies Definition 3; see also [22]).

**Definition 6.** *The addition and the scalar multiplication on $\mathfrak{F}(\mathbb{R})$ are defined as follows. For $A, B \in \mathfrak{F}(\mathbb{R})$ and $\lambda \geq 0$,*

$$\mu_{A \oplus B}(u) := \sup_{\{x,y \in \mathbb{R}: x+y=u\}} \min\{\mu_A(x), \mu_B(y)\}, \tag{3}$$

*for all $u \in \mathbb{R}$, and*

$$\lambda A(u) := \begin{cases} \mu_A(\frac{u}{\lambda}), & \lambda > 0 \\ \\ I_{\{0\}}(u), & \lambda = 0, \end{cases}$$

*for all $u \in \mathbb{R}$, where $I_{\{0\}}$ is the indicator function of $\{0\}$.*

A direct consequence of the previous definition is the following result.

**Lemma 1.** *If $A = (\gamma_1, \gamma_2, \gamma_3, \gamma_4)$ and $B = (\eta_1, \eta_2, \eta_3, \eta_4)$ are two trapezoidal fuzzy numbers, then*

*a) $A \oplus B = (\gamma_1 + \eta_1, \gamma_2 + \eta_2, \gamma_3 + \eta_3, \gamma_4 + \eta_4)$;*
*b) $\lambda A = (\lambda\gamma_1, \lambda\gamma_2, \lambda\gamma_3, \lambda\gamma_4)$, for each $\lambda \geq 0$.*

## 2.2 Order, Metric and Convergence

Let $D$ denote the set of all closed bounded intervals $A = [a_l, a_u]$ on the real line $\mathbb{R}$. For $A, B \in D$, with $A = [a_l, a_u]$, $B = [b_l, b_u]$ define

$$d(A, B) = \max(|a_l - b_l|, |a_u - b_u|).$$

It is not difficult to check that $d$ defines a metric on $D$, and that $(D, d)$ is a complete metric space.

Furthermore, for $A, B \in D$ define: $A \lesssim B$ if and only if $a_l \leq b_l$ and $a_u \leq b_u$. Note that "$\lesssim$" is a partial order on $D$.

Now, define $\rho : \mathfrak{F}(\mathbb{R}) \times \mathfrak{F}(\mathbb{R}) \longrightarrow \mathbb{R}$ by

$$\rho(\mu, \nu) = sup_{\alpha \in [0,1]} d(\mu_\alpha, \nu_\alpha), \tag{4}$$

with $\mu, \nu \in \mathfrak{F}(\mathbb{R})$. It is straightforward to see that $\rho$ is a metric in $\mathfrak{F}(\mathbb{R})$ [16].

Besides, for $\mu, \nu \in \mathfrak{F}(\mathbb{R})$ define

$$\mu \leq^* \nu \; if \; and \; only \; if \; \mu_\alpha \lesssim \nu_\alpha, \tag{5}$$

with $\alpha \in [0, 1]$.

*Remark 3.* Note that the binary relation "$\leq^*$" satisfies the axioms of a partial order relation on $\mathfrak{F}(\mathbb{R})$ and is called the *fuzzy max order* [11]. Moreover, if $\tilde{x}$ satisfies that $x \leq^* \tilde{x}$ for each $x \in \mathfrak{F}(\mathbb{R})$, then $\tilde{x}$ is an *upper bound* for $\mathfrak{F}(\mathbb{R})$. If the set of upper bounds of $\mathfrak{F}(\mathbb{R})$ has a least element, then this element is called the *supremum* of $\mathfrak{F}(\mathbb{R})$ [2, 26].

The proof of the following result can be consulted in [20].

**Lemma 2.** *The metric space* $(\mathfrak{F}(\mathbb{R}), \rho)$ *is complete.*

The following result is an extension of Lemma 1 and its proof is direct.

**Lemma 3.** *For trapezoidal fuzzy numbers the following statements hold:*

a) *If* $\{(\gamma_1^n, \gamma_2^n, \gamma_3^n, \gamma_4^n) : 1 \leq n \leq N\}$ *where $N$ is a positive integer, then*

$$\bigoplus_{n=1}^{N} (\gamma_1^n, \gamma_2^n, \gamma_3^n, \gamma_4^n) = (\sum_{n=1}^{N} \gamma_1^n, \sum_{n=1}^{N} \gamma_2^n, \sum_{n=1}^{N} \gamma_3^n, \sum_{n=1}^{N} \gamma_4^n).$$

b) *If* $u_n = \{(\gamma_1^n, \gamma_2^n, \gamma_3^n, \gamma_4^n) : n \geq 1\}$ *and* $\sum_{n=1}^{\infty} \gamma_i^n < \infty$, $i \in \{1, 2, 3, 4\}$, *then* $S_n := \bigoplus_{m=1}^{n} u_m$, $n \geq 1$, *converges to the trapezoidal fuzzy number* $(\sum_{n=1}^{\infty} \gamma_1^n, \sum_{n=1}^{\infty} \gamma_2^n, \sum_{n=1}^{\infty} \gamma_3^n, \sum_{n=1}^{\infty} \gamma_4^n)$.

## 2.3 Fuzzy Random Variables

Now, a fuzzy random variable will be defined. In this case, the definition proposed in [20] will be adopted.

**Definition 7.** *Let $(\Omega, \mathcal{F})$ be a measurable space and $(\mathbb{R}, \mathcal{B}(\mathbb{R}))$ be the measurable space of the real numbers endowed with the corresponding Borel $\sigma$-algebra $\mathcal{B}(\mathbb{R})$. A fuzzy random variable is a function $\tilde{X} : \Omega \longrightarrow \mathfrak{F}(\mathbb{R})$ such that for all $(\alpha, B) \in [0, 1] \times \mathcal{B}(\mathbb{R})$, $\{\omega \in \Omega : \tilde{X}_\alpha(\omega) \cap B \neq \varnothing\} \in \mathcal{F}$, where*

$$\tilde{X}_\alpha(\omega) = \left\{ x \in \mathbb{R} : \tilde{X}(\omega)(x) \geq \alpha \right\}.$$

**Definition 8.** *Let $(\Omega, \mathcal{F}, P)$ be a probability space and let $\tilde{X}$ be a discrete fuzzy random variable with the range $\{\tilde{s}_1, \tilde{s}_2, ..., \tilde{s}_l\} \subseteq \mathfrak{F}(\mathbb{R})$. The mathematical expectation of $\tilde{X}$ is a fuzzy number, $E^*[\tilde{X}]$, such that*

$$E^*[\tilde{X}] = \bigoplus_{i=1}^{l} \tilde{s}_i P(\tilde{X} = \tilde{s}_i). \tag{6}$$

A proof of the following result should be consulted in [12].

**Lemma 4.** *Let $\tilde{X}$ and $\tilde{Y}$ be discrete fuzzy random variables with a finite range. Then*

*a) $E^*[\tilde{X}] \in \mathfrak{F}(\mathbb{R})$.*
*b) $E^*[\tilde{X} + \tilde{Y}] = E^*[\tilde{X}] + E^*[\tilde{Y}]$.*
*c) $E^*[\lambda \tilde{X}] = \lambda E^*[\tilde{X}], \lambda \geq 0$.*

## 3  Discounted MDPs

In this section, the basic theory of Markov Decision Processes (MDPs) will be presented, which will be necessary to introduce an adequate fuzzy MDP, in the next section. Detailed literature on the theory of Markov decision processes can be consulted in the references: [14] and [21].

**Definition 9.** *A Markov decision model is a five-tuple which consists of the following elements:*

$$M := (X, A, \{A(x) : x \in X\}, Q, R), \tag{7}$$

*where*

*a) $X$ is a finite set, which is called the* state space.
*b) $A$ is a Borel space, $A$ is denominated the* control *or* action space.
*c) $\{A(x) : x \in X\}$ is a family of nonempty subsets $A(x)$ of $A$, whose elements are the* feasible actions *when the state of system is $x$.*
*c) $Q$ is the* transition law, *which is a stochastic kernel on $X$ given $\mathbb{K} := \{(x, a) : x \in X, a \in A(x)\}$, $\mathbb{K}$ is denominated the set of feasible state-actions pairs.*
*d) $R : \mathbb{K} \longrightarrow \mathbb{R}$ is the* one-step *reward function.*

Now, given a Markov control Model $M$, the concept of policy will be introduced. A *policy* is a sequence $\pi = \{\pi_t : t = 0, 1, ...\}$ of stochastic kernels $\pi_t$ on the control set $A$ given the history $\mathbb{H}_t$ of the process up to time $t$, where $\mathbb{H}_t := \mathbb{K} \times \mathbb{H}_{t-1}, t = 1, 2, ...$ and $\mathbb{H}_0 = X$. The set of all policies will be denoted by $\Pi$. A *deterministic Markov policy* is a sequence $\pi = \{f_t\}$ such that $f_t \in \mathbb{F}$, for $t = 0, 1, ...$, where $\mathbb{F}$ denotes the set of functions $f : X \longrightarrow A$ such that $f(x) \in A(x)$, for all $x \in X$. The set of all Markovian policies will be denoted by $\mathbb{M}$. A Markov policy $\pi = \{f_t\}$ is said to be *stationary* if $f_t$ is independent of $t$, i.e., $f_t = f \in \mathbb{F}$, for all $t = 0, 1, ....$ In this case, $\pi$ is identified with $f$ and $\mathbb{F}$ is denominated the set of *stationary policies*.

Let $(\Omega', \mathcal{F}')$ be the measurable space consisting of the canonical sample space $\Omega' = \mathbb{H}_\infty := (X \times A)^\infty$ and $\mathcal{F}'$ be the corresponding product $\sigma$-algebra. The elements of $\Omega'$ are sequences of the form $\omega = (x_0, a_0, x_1, a_1, ...)$ with $x_t \in X$ and $a_t \in A$ for all $t = 0, 1, ....$ The projections $x_t$ and $a_t$ from $\Omega'$ to the sets $X$ and $A$ are called state and action variables, respectively.

Let $\pi = \{\pi_t\}$ be an arbitrary policy and $\mu$ be an arbitrary probability measure on $X$ called the initial distribution. Then, by the theorem of C. Ionescu-Tulcea [14], there is a unique probability measure $P_\mu^\pi$ on $(\Omega', \mathcal{F}')$ which is supported on $\mathbb{H}_\infty$, i.e., $P_\mu^\pi(\mathbb{H}_\infty) = 1$. The stochastic process $(\Omega', \mathcal{F}', P_\mu^\pi, \{x_t\})$ is called a discrete-time Markov control process or a Markov decision process.

The expectation operator with respect to $P_\mu^\pi$ is denoted by $E_{\mu,\pi}$. If $\mu$ is concentrated at the initial state $x \in X$, then $P_\mu^\pi$ and $E_{\mu,\pi}$ are written as $P_x^\pi$ and $E_{x,\pi}$, respectively.

The transition law of a Markov control process (see (7)) is often specified by a difference equation of the form

$$x_{t+1} = F(x_t, a_t, \xi_t), \tag{8}$$

$t = 0, 1, 2, ...$, with $x_0 = x \in X$ known, where $\{\xi_t\}$ is a sequence of independent and identically distributed (i.i.d.) random variables with values in a finite space $S$ and a common distribution $\Delta$, independent of the initial state $x_0$. In this case, the transition law $Q$ is given by

$$Q(B|x, a) = E[I_B(F(x, a, \xi))],$$

$B \subseteq X$, $(x, a) \in K$, $E$ is the expectation with respect to distribution $\Delta$, $\xi$ is a generic element of the sequence $\{\xi_t\}$ and $I_B(\cdot)$ denotes the indicator function of the set $B$.

**Definition 10.** *Let $(X, A, \{A(x) : x \in X\}, Q, R)$ be a Markov model, then the expected total discounted reward is defined as follows:*

$$v(\pi, x) := E_{x,\pi}\left[\sum_{t=0}^{\infty} \beta^t R(x_t, a_t)\right], \tag{9}$$

$\pi \in \Pi$ *and* $x \in X$, *where* $\beta \in (0, 1)$ *is a given discount factor. Furthermore, the expected total discounted reward with a finite horizon is defined as follows:*

$$v_T(\pi, x) := E_{x,\pi}\left[\sum_{t=0}^{T-1} \beta^t R(x_t, a_t)\right], \tag{10}$$

*for each* $x \in X$ *and* $\pi \in \Pi$, *where* $T$ *is a positive integer.*

**Definition 11.** *The* optimal value function *is defined as*

$$V(x) := sup_{\pi \in \Pi} v(\pi, x), \tag{11}$$

$x \in X$. *Then the* optimal control problem *is to find a policy* $\pi^* \in \Pi$ *such that*

$$v(\pi^*, x) = V(x),$$

$x \in X$, *in which case,* $\pi^*$ *is said to be the* optimal policy. *Similar definitions can be established analogously for* $v_T$. *In this case,* $V_T$ *denotes the optimal value function for the optimal control problem with a finite horizon.*

**Assumption 1.** *a) For each* $x \in X$, $A(x)$ *is a compact set on* $\mathcal{B}(A)$, *where* $\mathcal{B}(A)$ *is the Borel* $\sigma$-*algebra of space* $A$.
*b) The reward function* $R$ *is a non-negative and bounded function.*
*c) For every* $x, y \in X$, *the mappings* $a \mapsto R(x, a)$ *and* $a \mapsto Q(\{y\}|x, a)$ *are continuous in* $a \in A(x)$.

The proof of the following theorem which provides the Dynamic Programming approach can be consulted in [14] and [21].

**Theorem 2.** *Under Assumption 1, the following statements hold:*

*a) Define* $W_T(x) = 0$ *and for each* $n = T - 1, ..., 1, 0$, *consider*

$$W_n(x) = \max_{a \in A(x)} \{R(x, a) + \beta E[W_{n+1}(F(x, a, \xi))]\}. \tag{12}$$

$x \in X$. *Then for each* $n = 0, 1, ..., T - 1$, *there exists* $f_n \in \mathbb{F}$ *such that*

$$W_n(x) = R(x, f_n(x)) + \beta E[W_{n+1}(F(x, f_n(x), \xi))],$$

$x \in X$. *In this case,* $\pi^* = \{f_0, ..., f_{T-1}\} \in \mathbb{M}$ *is the optimal policy and* $V_T(x) = v_T(\pi^*, x) = W_0(x)$, $x \in X$.
*b) The optimal value function* $V$, *satisfies the following Dynamic Programming Equation:*

$$V(x) = \max_{a \in A(x)} \{R(x, a) + \beta E[V(F(x, a, \xi))]\}, \tag{13}$$

$x \in X$.
*c) There exists a policy* $f^* \in \mathbb{F}$ *such that the control* $f^*(x) \in A(x)$ *attains the maximum in (13), i.e. for all* $x \in X$,

$$V(x) = R(x, f^*(x)) + \beta E[V(F(x, f^*(x), \xi))]. \tag{14}$$

*d) Define the* value iteration functions *as follows:*

$$V_n(x) = \min_{a \in A(x)} \{c(x, a) + \beta E[V_{n-1}(F(x, f^*(x), \xi))]\}, \tag{15}$$

*for all* $x \in X$ *and* $n = 1, 2, ...,$ *with* $V_0(\cdot) = 0$. *Then the sequence* $\{V_n\}$ *of the value iteration functions converges pointwise to the optimal value function* $V$, *i.e.*

$$\lim_{n \to \infty} V_n(x) = V(x),$$

$x \in X$.

*Remark 4.* As a consequence of Theorem 2, the following facts hold:

a) From Theorem 2(a), in the case of the discounted expected reward with a finite horizon the optimum is attained in a Markovian policy, hence

$$sup_{\pi \in \Pi} v_T(\pi, x) = sup_{\pi \in \mathbb{M}} v_T(\pi, x),$$

$x \in X$.

b) From Theorem 2(c), in the case of discounted expected reward with an infinite horizon the optimum is reached in a stationary policy. Then, it follows that

$$sup_{\pi \in \Pi} v(\pi, x) = sup_{f \in \mathbb{F}} v(f, x),$$

$x \in X$.

## 4  Discounted MDPs with Fuzzy Rewards

Consider a Markov decision model $\tilde{M} = (X, A, \{A(x) : x \in X\}, Q, \tilde{R})$, where the first four components are the same as in the model given in (7). The fifth component, $\tilde{R}$, corresponds to a fuzzy reward function on $\mathbb{K}$.

The evolution of a stochastic fuzzy system is as follows: if the system is in the state $x_t = x \in X$ at time $t$ and the control $a_t = a \in A(x)$ is applied, then two things happen:

a) a fuzzy reward $\tilde{R}(x, a)$ is obtained.
b) the system jumps to the next state $x_{t+1}$ according to the transition law $Q$, i.e.

$$Q(B|x, a) = Prob(x_{t+1} \in B | x_t = x, a_t = a),$$

with $B \subseteq X$.

For each policy $\pi \in \mathbb{M}$ and state $x \in X$, let

$$\tilde{v}(i, \pi) = \bigoplus_{t=0}^{T-1} \beta^t E^*_{i,\pi} \left[ \tilde{R}(x_t, a_t) \right], \tag{16}$$

where $T$ is a positive integer and $E^*_{i,\pi}$ is the expectation with respect to $\tilde{P}^\pi_x$ which is defined by (6). The expression given in (16) is called the expected total discounted fuzzy reward with a finite horizon. Furthermore, the following objective function will be considered:

$$\tilde{V}(i, \pi) = \bigoplus_{t=0}^{\infty} \beta^t E^*_{i,\pi} \left[ \tilde{R}(x_t, a_t) \right], \tag{17}$$

and, the expectation in (17) is defined by (6), when $\{a_t\}$ is induced by a stationary policy $\pi$.

In this way, the control problem of interest is the maximization of the finite or infinite horizon expected total discounted fuzzy reward (see (16) and (17), respectively). The following assumption is considered for the reward function of fuzzy model $\tilde{M}$.

**Assumption 3.** Let $\gamma_1, \gamma_2, \gamma_3$ and $\gamma_4$ be real numbers, such that $0 < \gamma_1 < \gamma_2 \leq \gamma_3 < \gamma_4$. It will be assumed that the fuzzy reward is a trapezoidal fuzzy number (see Definition 4), specifically

$$\tilde{R}(x, a) = R(x, a)(\gamma_1, \gamma_2, \gamma_3, \gamma_4) \tag{18}$$

for each $(x, a) \in \mathbb{K}$, where $R : \mathbb{K} \longrightarrow \mathbb{R}$ is the reward function of the model introduced in Sect. 3.

*Remark 5.* Observe that, under Assumption 3 and Lemma 3, the fuzzy reward (16) is a trapezoidal fuzzy number.

## 4.1   Fuzzy Optimal Control Problems

In this section, results of the convergence of the fuzzy reward (16) to the infinite horizon expected total discounted fuzzy reward (17) will be presented, when $T$ goes to infinity. Later, the existence of optimal policies and validity of dynamic programming will be verified.

**Lemma 5.** *Suppose that Assumption 3 holds. Then, for each $i \in X$, $\pi \in \mathbb{F}$ (see Remark 4), $\{\tilde{V}_T(\pi, i) : T = 0, 1, ...\}$ converges and*

$$\tilde{v}(i, \pi) = lim_{T \to \infty} \tilde{v}_T(\pi, i) = \bigoplus_{t=0}^{\infty} \eta^t E_{i,\pi}^* \left[ \tilde{R}(x_t, a_t) \right] = v(i, \pi)(\gamma_1, \gamma_2, \gamma_3, \gamma_4),$$

*where $v(i, \pi) = \sum_{t=0}^{\infty} \eta^t E_{i,\pi} [R(x_t, a_t)] \in \mathbb{R}$.*

*Proof.* Let $\pi \in \Pi$ and $x \in X$ be fixed. To simplify the notation in this proof, it will be denoted that $v = v(\pi, x)$ and $v_T = v_T(\pi, x)$ (see (9) and (10)). Then, the $\alpha$-cut of (16) is given by

$$\Delta^T := (\gamma_1 v_T, \gamma_2 v_T, \gamma_3 v_T, \gamma_4 v_T)_\alpha$$
$$= [\gamma_1(1 - \alpha)v_T + \alpha\gamma_2 v_T, \gamma_4(1 - \alpha)v_T + \alpha\gamma_3 v_T].$$

Analogously,

$$\Delta := (\gamma_1 v, \gamma_2 v, \gamma_3 v, \gamma_4 v)_\alpha$$
$$= [\gamma_1(1 - \alpha)v + \alpha\gamma_2 v, \gamma_4(1 - \alpha)v + \alpha\gamma_3 v].$$

Hence, by (4), it is obtained that

$$d(\Delta^T, \Delta) = sup_{\alpha \in [0,1]} d(\Delta^T, \Delta).$$

Now, due to the identity $\max(c, b) = (c + b + |b - c|)/2$ with $b, c \in \mathbb{R}$, it yields that

$$d(\Delta^T, \Delta) = (1 - \alpha)\gamma_3(v - v_T) + \alpha\gamma_2(v - v_T).$$

Then,

$$d(\Delta_T, \Delta) = sup_{\alpha \in [0,1]}(v - v_T)(\gamma_3 - \alpha(\gamma_3 - \gamma_2)) \tag{19}$$
$$= (v - v_T)D.$$

Therefore, when $T$ goes to infinity in (19), it concludes that

$$\lim_{T \longrightarrow \infty} \rho(\tilde{v}_T, \tilde{v}) = \lim_{T \longrightarrow \infty} (v - v_T)\gamma_3$$
$$= 0.$$

The second equality is a consequence of (9) and (10).

**Definition 12.** *The infinite-horizon fuzzy optimal control problem consists in determining a policy $\pi^* \in \mathbb{F}$ such that*

$$\tilde{v}(\pi, x) \leq^* \tilde{v}(\pi^*, x),$$

*for all $\pi \in \mathbb{F}$ and $x \in X$. In consequence (see Remark 4 (b)),*

$$\tilde{v}(\pi^*, x) = sup_{\pi \in \mathbb{F}}\tilde{v}(\pi, x),$$

*for all $x \in X$ (see Remark 3). In this case, the optimal fuzzy value function is defined as follows:*

$$\hat{V}(x) = \tilde{v}(\pi^*, x),$$

*$x \in X$ and $\pi^*$ is called the optimal policy for the fuzzy optimal control problem.*

*Remark 6.* Similar definitions can be stated for $\tilde{v}_T$, the expected total discounted fuzzy reward with a finite horizon $T$. In this case the optimal fuzzy value is denoted by $\hat{V}_T$, and (see Remark 4(a)),

$$\hat{V}_T(x) = \tilde{v}_T(\pi^*, x) = sup_{\pi \in M}\tilde{v}_T(\pi, x),$$

for all $x \in X$, of course, if such $\pi^*$ exists then it is called the optimal policy for the fuzzy optimal control problem with a horizon $T$.

A direct consequence of Definition 12, Remark 6, and Theorem 2 is the next result.

**Theorem 4.** *Under Assumptions 1 and 3, the following statements hold:*

a) *The optimal policy $\pi^*$ of the crisp finite optimal control problem (see (10)) is the optimal policy for $\tilde{v}_T$, i.e. $\tilde{v}_T(\pi^*, x) = sup_{\pi \in M}\tilde{v}_T(\pi, x)$ for all $\pi \in \Pi$ and $x \in X$.*
b) *The optimal fuzzy value function is given by*

$$\hat{V}_T(x) = V_T(x)(\gamma_1, \gamma_2, \gamma_3, \gamma_4), \tag{20}$$

*$x \in X$, where $\hat{V}_T(x) = sup_{\pi \in M}\tilde{v}_T(\pi, x)$, $x \in X$.*

*Proof.* a) Let $\pi \in M$ and $x \in X$ be fixed. Then, from (16), it is obtained that

$$\tilde{v}_t(\pi, x) = v_T(\pi, x)(\gamma_1, \gamma_2, \gamma_3, \gamma_4),$$

where Assumption 3 and Lemma 3 were applied. Now, observe that the $\alpha$-cut of $\tilde{v}_T(\pi, x)$ is given by the following closed interval:

$$\tilde{v}_t(\pi, x)_\alpha = [\gamma_1 v_T(\pi, x) + \alpha(\gamma_2 - \gamma_1)v_T(\pi, x), \gamma_4 v_T(\pi, x) - \alpha v_T(\pi, x)(\gamma_4 - \gamma_3)].$$

On the other hand, by Theorem 2, there exists an optimal policy $\pi^* \in \mathbb{M}$ such that, $v_T(\pi, x) \leq v_T(\pi^*, x)$. Then, observe that the extremes of the $\tilde{v}_t(\pi, x)_\alpha$ satisfy the following inequalities:

$$\gamma_1 v_T(\pi, x) + \alpha(\gamma_2 - \gamma_1)v_T(\pi, x) \leq \gamma_1 v_T(\pi^*, x) + \alpha(\gamma_2 - \gamma_1)v_T(\pi^*, x)$$
$$\gamma_4 v_T(\pi, x) - \alpha v_T(\pi^*, x)(\gamma_4 - \gamma_3) \leq \gamma_4 v_T(\pi^*, x) - \alpha v_T(\pi^*, x)(\gamma_4 - \gamma_3).$$

Consequently, $\tilde{v}_T(\pi, x) \leq^* \tilde{v}_T(\pi^*, x)$. Since $x \in X$ and $\pi \in \Pi$ are arbitrary, the result follows, due to Definition 12.

b) By Theorem 5a), it follows that

$$\tilde{V}_T(x) = v(\pi^*, x)(\gamma_1, \gamma_2, \gamma_3, \gamma_4),$$

for each $x \in X$, thus applying Theorem 2, it is concluded that

$$\tilde{V}_T(x) = V_T(x)(\gamma_1, \gamma_2, \gamma_3, \gamma_4),$$

$x \in X$.

The proof of Theorem 5 is similar to proof of Theorem 4, for this reason the proof is omitted.

**Theorem 5.** *Under Assumptions 1 and 3, the following statements hold:*

a) *The optimal policy of the fuzzy control problem is the same as the optimal policy of the optimal control problem.*
b) *The optimal fuzzy value function is given by*

$$\tilde{V}(x) = V(x)(\gamma_1, \gamma_2, \gamma_3, \gamma_4), x \in X. \tag{21}$$

In the following sections, Theorems 4 and 5 will be illustrated in several examples.

## 5 Examples

### 5.1 A Fuzzy Inventory Control System

In this section, first a classical example of inventory control system [21] will be presented, later a trapezoidal fuzzy inventory control system will be introduced. The optimal solution of the fuzzy inventory is obtained by an application of Theorem 4 and the solution of the crisp inventory system.

The next example is addressed in [21], below there is a summary of the points of interest to introduce its fuzzy version. Consider the following situation: a warehouse where every certain period of time the manager carries out an inventory to determine the quantity of product stored. Based on such information, a decision is made whether or not to order a certain amount of additional product from a supplier. The manager's goal is to maximize the profit obtained. The demand for the product is assumed to be random, known probability distribution. The following assumptions will be treated to propose the mathematical model.

## Inventory Assumptions

a) The decision of an additional order is made at the beginning of the period and is delivered immediately.
b) Product demands are received throughout the period of time but are fulfilled in the last instant of the time of the period.
c) There are no unfilled orders.
c) Revenues and the distribution of demand do not vary with the period.
d) The product is only sold in whole units.
e) The warehouse has a capacity for $M$ units, where $M$ is a positive integer.

Then, under previous assumption, the state space is given by $X := \{0, 1, 2, ..., M\}$, the action space and admissible action set are given by $A := \{0, 1, 2, ...\}$ and $A(x) := \{0, 1, 2, ..., M - x\}$, $x \in X$, respectively.

Now, consider the following variables: let $x_t$ denote the inventory at time $t = 0, 1, ...$, the evolution of the system is modeled by a dynamics that follows a Lindley process [3]:

$$x_{t+1} = (x_t + a_t - D_{t+1})^+, \tag{22}$$

with $x_0 = x \in X$ known, where $(z)^+ = max\{0, z\}$, $z \in \mathbb{R}$, and

a) $a_t$ denotes the control or decision applied in the instant $t$ and it represents the quantity ordered by the inventory manager (or decision maker).
b) The sequence $\{D_t\}$ is conformed by independent and identically distributed non-negative random variables with common distribution $p_j := \mathbb{P}(D = j), j = 0, 1, ...,$ where $D_t$ denotes the demand within the period of time $t$.

Observe that the difference equation given in (22) induces a stochastic kernel defined on $X$ given $\mathbb{K} := \{(x, a) : x \in X, a \in A(x)\}$, as follows

$$Q(x_{t+1} \in (-\infty, y]) | x_t = x, a_t = a) = 1 - \Delta(x + a - y),$$

where $\Delta$ is the distribution of $D$ with $x \in X$, $y, a \in \{0, 1, ...\}$ and $Q(x_{t+1} \in (-\infty, y]) | x_t = x, a_t = a) = 0$, if $x \in X, a \in \{0, 1, ...\}$ and $y < 0$. Then it follows that

$$Q(\{x_{t+1} = y\} | x, a) = \begin{cases} 0 & if \quad M \geq y > x + a \\ p_{x+a-y} & if \; M \geq x + a \geq y > 0 \\ q_{x+a} & if \quad M \geq x + a, y = 0 \end{cases}$$

The step reward function is given by $R(x, a) = E[H(x + a - (x + a - D)^+)]$, $(x, a) \in \mathbb{K}$, where $H : \{0, 1, ...\} \to \{0, 1, ...\}$ is the revenue function, which is a known function and $D$ is a generic element of the sequence $\{D_t\}$. Equivalently, $R(x, a) = F(x + a)$, $(x, a) \in \mathbb{K}$, where

$$F(u) := \sum_{k=0}^{u-1} H(k)p_k + H(u)q_u, \tag{23}$$

with $q_u := \sum_{k=u}^{\infty} p_k$. The objective in this section is to maximize the total discounted reward with a finite horizon, see (16).

In particular, suppose that the horizon is $T = 156$, the state space $X = \{0, 1, ..., 9\}$, the revenue function $H(u) = 5u$ and the transition law is given in Fig. 2.

| | [,1] | [,2] | [,3] | [,4] | [,5] | [,6] | [,7] | [,8] | [,9] | [,10] |
|---|---|---|---|---|---|---|---|---|---|---|
| [1,] | 1.0000000 | 0.00000000 | 0.00000000 | 0.00000000 | 0.00000000 | 0.00000000 | 0.00000000 | 0.00000000 | 0.00000000 | 0.00000000 |
| [2,] | 1.0000000 | 0.00000000 | 0.00000000 | 0.00000000 | 0.00000000 | 0.00000000 | 0.00000000 | 0.00000000 | 0.00000000 | 0.00000000 |
| [3,] | 0.9777778 | 0.02222222 | 0.00000000 | 0.00000000 | 0.00000000 | 0.00000000 | 0.00000000 | 0.00000000 | 0.00000000 | 0.00000000 |
| [4,] | 0.9333333 | 0.04444444 | 0.02222222 | 0.00000000 | 0.00000000 | 0.00000000 | 0.00000000 | 0.00000000 | 0.00000000 | 0.00000000 |
| [5,] | 0.8666667 | 0.06666667 | 0.04444444 | 0.02222222 | 0.00000000 | 0.00000000 | 0.00000000 | 0.00000000 | 0.00000000 | 0.00000000 |
| [6,] | 0.7777778 | 0.08888889 | 0.06666667 | 0.04444444 | 0.02222222 | 0.00000000 | 0.00000000 | 0.00000000 | 0.00000000 | 0.00000000 |
| [7,] | 0.6666667 | 0.11111111 | 0.08888889 | 0.06666667 | 0.04444444 | 0.02222222 | 0.00000000 | 0.00000000 | 0.00000000 | 0.00000000 |
| [8,] | 0.5333333 | 0.13333333 | 0.11111111 | 0.08888889 | 0.06666667 | 0.04444444 | 0.02222222 | 0.00000000 | 0.00000000 | 0.00000000 |
| [9,] | 0.3777778 | 0.15555556 | 0.13333333 | 0.11111111 | 0.08888889 | 0.06666667 | 0.04444444 | 0.02222222 | 0.00000000 | 0.00000000 |
| [10,] | 0.2000000 | 0.17777778 | 0.15555556 | 0.13333333 | 0.11111111 | 0.08888889 | 0.06666667 | 0.04444444 | 0.02222222 | 0.00000000 |

**Fig. 2.** Transition law.

In consequence, the output of the program is obtained as illustrated in Fig. 3. In this matrix, the last column represents the optimal policy and the penultimate column, the value function, for each state $x \in \{0, 1, ..., 9\}$. The other input of the matrix represents the following:

$$G(x, a) := R(x, a) + \alpha E[W_1(F(x, a, D))],$$

$(x, a) \in \mathbb{K}$.

| | | | | | | | | | | $V_t$ | $\pi^*$ |
|---|---|---|---|---|---|---|---|---|---|---|---|
| [1,] | 285.0000 | 290.0000 | 294.8889 | 299.5555 | 303.8889 | 307.7778 | 311.1111 | 313.7778 | 315.6666 | 316.6666 | 316.6666 | 9 |
| [2,] | 290.0000 | 294.8889 | 299.5555 | 303.8889 | 307.7778 | 311.1111 | 313.7778 | 315.6666 | 316.6666 | 0.0000 | 316.6666 | 8 |
| [3,] | 294.8889 | 299.5555 | 303.8889 | 307.7778 | 311.1111 | 313.7778 | 315.6666 | 316.6666 | 0.0000 | 0.0000 | 316.6666 | 7 |
| [4,] | 299.5555 | 303.8889 | 307.7778 | 311.1111 | 313.7778 | 315.6666 | 316.6666 | 0.0000 | 0.0000 | 0.0000 | 316.6666 | 6 |
| [5,] | 303.8889 | 307.7778 | 311.1111 | 313.7778 | 315.6666 | 316.6666 | 0.0000 | 0.0000 | 0.0000 | 0.0000 | 316.6666 | 5 |
| [6,] | 307.7778 | 311.1111 | 313.7778 | 315.6666 | 316.6666 | 0.0000 | 0.0000 | 0.0000 | 0.0000 | 0.0000 | 316.6666 | 4 |
| [7,] | 311.1111 | 313.7778 | 315.6666 | 316.6666 | 0.0000 | 0.0000 | 0.0000 | 0.0000 | 0.0000 | 0.0000 | 316.6666 | 3 |
| [8,] | 313.7778 | 315.6666 | 316.6666 | 0.0000 | 0.0000 | 0.0000 | 0.0000 | 0.0000 | 0.0000 | 0.0000 | 316.6666 | 2 |
| [9,] | 315.6666 | 316.6666 | 0.0000 | 0.0000 | 0.0000 | 0.0000 | 0.0000 | 0.0000 | 0.0000 | 0.0000 | 316.6666 | 1 |
| 10,] | 316.6666 | 0.0000 | 0.0000 | 0.0000 | 0.0000 | 0.0000 | 0.0000 | 0.0000 | 0.0000 | 0.0000 | 316.6666 | 0 |

**Fig. 3.** Optimal value function $V_T$ and optimal policy $\pi^*$.

In conclusion, the optimal value function is $V_T(x) = 316.6666$ for each $x \in X$ and the optimal policy is given by $f_t(x) = M - x, t = 0, 1, ..., T - 1, x \in X$ with $M = 9$.

Now, considering that in operations research it is often difficult for a manager to control inventory systems, due to the fact that data in each stage of observation is not always certain, then a fuzziness approach should be applied. In this way, take into account the previous inventory system in a fuzzy environment, that is, the reward function given in Assumption 3 will be considered:

$$\tilde{R}(x, a) = (\gamma_1 R(x, a), \gamma_2 R(x, a), \gamma_3 R(x, a), \gamma_4 R(x, a)),$$

with $0 < \gamma_1 < \gamma_2 \le \gamma_3 < \gamma_4$. Then, by Theorem 4, it follows that the optimal policy of the fuzzy optimal control problem is given by $\tilde{\pi}^* = \{f_0, ..., f_{T-1}\}$, where $f_t(x) = M - x, t = 0, 1, ..., T - 1, x \in X$ and the optimal value function is given by

$$\tilde{V}_T(x) = V_T(x)(\gamma_1, \gamma_2, \gamma_3, \gamma_4),$$

$x \in X$.

---

**Algorithm 1.** To calculate the optimal value and optimal policy.

---

**Input:** MDP
**Output:** The optimal value vector.
An optimal policy
**Initialize** $W_T(x, A) = 0$, $W_T^*(x) = 0$,
$K_T(x) = W_T^*(x)$.
$t = T - 1$
**repeat**

    **for** $x \in S$ **do**
      $f_x = 0$
      $a(x) = f_x$
      $W(x, a(x)) = R(x, a(x)) +$
                  $\beta \sum_{i=0}^{Z} Q(y|x + a(x)) W_{t+1}(y, 0)$

      $A(x) = 1, ..., M - x$
      **for** $a \in A(x)$ **do**
         $W_t(x, a) = R(x, a) +$
                  $\beta \sum_{y=0}^{Z} Q(y|x + a) W_{t+1}(y, 0)$
        **if** $W_t(x, a) \geq W(x, a(x))$   **do**
           $W(x, a(x)) = W_t(x, a)$
           $f_x = a$
      **end for**

      $W_t(x) = W_t(x, f_x)$

      $W_t(x, 0) = W_t(x)$

      **if** $W_t(x) \geq K_{t+1}(x)$   **do**
        $K_t(x) = W_t(x)$

      $W^*(x) = K_t(x)$
    **end for**

    $t = n - 1$
**until** $t = 0$

---

## 5.2   A Portfolio Choice Problem

Let $X = \{\chi_0, \chi_1\}$, $0 < \chi_0 < \chi_1$, $A(\chi) = [0, 1]$, $\chi \in X$. The transition law is given by

$$Q(\{\chi_0\}|\chi_0, a) = p, \tag{24}$$
$$Q(\{\chi_1\}|\chi_0, a) = 1 - p, \tag{25}$$
$$Q(\{\chi_1\}|\chi_1, a) = q, \tag{26}$$

$$Q(\{\chi_0\}|\chi_1, a) = 1 - q, \tag{27}$$

for all $a \in [0, 1]$, where $0 \le p \le 1$ and $0 \le q \le 1$. The reward is given by a function $R(\chi, a)$, $(\chi, a) \in \mathbb{K}$ that met:

**Assumption 6.** *(a)* $R$ depends only of $a$, that is $R(\chi, a) = U(a)$, for all $(\chi, a) \in \mathbb{K}$, where $U$ is non-negative and continuous.
*(b)* There is $a^* \in [0, 1]$ such that

$$max_{a \in [0,1]} U(a) = U(a^*),$$

for all $\chi \in X$.

An interpretation of this example is given in the following remark.

*Remark 7.* The states $\chi_0$ and $\chi_1$ could represent the behavior of certain stock market, which is bad ($\equiv \chi_0$) and good ($\equiv \chi_1$). It is assumed that, for each $a$ and $t = 0, 1, \cdots$, the probability of going from $\chi_0$ to $\chi_0$ is $p$ (resp. the probability of going from $\chi_0$ to $\chi_1$ is $1 - p$); moreover, for each $a$ and $t = 0, 1, \cdots$, the probability of going from $\chi_1$ to $\chi_1$ is $q$ (resp. the probability of going from $\chi_1$ to $\chi_0$ is $1 - q$). Now, specifically, suppose that in a dynamic portfolio choice problem, two assets are available to an investor. One is risky-free, and the risk-rate $r > 0$ is assumed to be known and constant over time. The other asset is risky with a stochastic return having mean $\mu$ and a variance $\sigma^2$. Following Example 1.24 in [27], the expected utility of the investor could be given for the expression:

$$U(a) = a\mu + (1 - a)r - \frac{k}{2}a^2\sigma^2, \tag{28}$$

where $a \in [0, 1]$ is the fraction of its money that the investor invests in the risky asset and the remainder $1 - a$, he/she invests in the riskless asset. In (28), $k$ represents the value that the investor places on the variance relative to the expectation. Observe that if $\mu > \dfrac{k\sigma^2}{2}$, then $U$ defined in (28) is positive in $[0, 1]$ (in fact, in this case $U(0) = r > 0$ and $U(1) = \mu - \dfrac{k\sigma^2}{2} > 0$); moreover, it is possible to prove (see [27]) that if $0 < \mu - r < k\sigma^2$, then $max_{a \in [0,1]} U(a)$ is attained for $a^* \in (0, 1)$ given by

$$a^* = \frac{\mu - r}{k\sigma^2}.$$

Hence, taking $R(\chi, a) = U(a)$, $\chi \in X$, and $a \in [0, 1]$, where $U$ is given by (28), and considering the last two inequalities given in the previous paragraph, Assumption 6 holds.

**Lemma 6.** *Suppose that Assumption 6 holds. Then, for Example 2,*

$$V(\chi) = \frac{U(a^*)}{1 - \beta}$$

*and* $f^*(\chi) = a^*$, *for all* $\chi \in X$.

*Proof.* Firstly, the value iteration functions will be found: $V_n$, for $n = 1, 2, \ldots$. By Theorem 2,

$$V_1(\chi_0) = \max_{a \in [0,1]} U(a),$$

this implies that $V_1(\chi_0) = U(a^*)$. In a similar way, it is possible to obtain that $V_1(\chi_1) = U(a^*)$.

Now, for $n = 2$,

$$
\begin{aligned}
V_2(\chi_0) &= \max_{a \in [0,1]} \{U(a) + \beta[V_1(\chi_1)(1 - p) + V_1(\chi_0)p]\} \\
&= U(a^*) + \beta[V_1(\chi_1)(1 - p) + V_1(\chi_0)p] \\
&= U(a^*) + \beta[U(a^*)(1 - p) + U(a^*)p] \\
&= U(a^*) + \beta U(a^*).
\end{aligned}
$$

Analogously, $V_2(\chi_1) = U(a^*) + \beta U(a^*)$. Continuing in this way, it is obtained that

$$V_n(\chi_0) = V_n(\chi_0) = U(a^*) + \beta U(a^*) + \ldots + \beta^{n-1} U(a^*),$$

for all $n = 1, 2, \ldots$.

By Theorem 2, $V_n(\chi) \to V(\chi)$, $n \to \infty$, $\chi \in X$, which implies that $V(\chi) = \dfrac{U(a^*)}{1 - \beta}$, $\chi \in X$. And, from the Dynamic Programming Equation (see (13)), it follows that $f^*(\chi) = a^*$, for all $\chi \in X$.

Now, suppose that the fuzzy reward function is given by

$$\tilde{R}(x, a) = (\gamma_1, \gamma_2, \gamma_3, \gamma_4) R(x, a),$$

with $(x, a) \in \mathbb{K}$. Then, as a consequence of Theorem 5 the following result is obtained.

**Lemma 7.** *For the fuzzy version of Portfolio Choice Problem, it results that* $\tilde{V}(\chi) = V(\chi)(\gamma_1, \gamma_2, \gamma_3, \gamma_4)$ *and* $f^*(\chi) = a^*$, *for all* $\chi \in X$.

## 5.3   A Two-Person Game

Now, a model of a stochastic game between two players who seek to maximize their total discounted rewards is presented. Let $J_1$ and $J_2$ denote the players/ investors. Each of them follows a decision model similar to the one proposed in Sect. 5.2. That is, $J_1$ has a decision model of the type: $(X, A, Q, R_1)$, where $X = \{\chi_0, \chi_1\}$, $0 < \chi_0 < \chi_1$, $A = A(\chi) = [0, 1]$, $\chi \in X$. The transition law $Q$ is given as in Sect. 5.2 (notice that, in fact, $Q$ given in (24–27) is independent of the decision $a$), and the reward is given by the function $R_1 = U_1$ with

$$U_1(a) = a\mu_1 + (1 - a)r_1 - \frac{k_1}{2}a^2\sigma_1^2,$$

where $a \in [0, 1]$. Moreover, it is assumed that $0 < \mu_1 - r_1 < k_1\sigma_1^2$, then $max_{a \in [0,1]} U_1(a)$ is attained for $a^* \in (0, 1)$ given by

$$a^* = \frac{\mu_1 - r_1}{k\sigma_1^2}.$$

Let $\mathbb{F}$ be the corresponding set of stationary strategies for $J_1$. Observe that

$$\mathbb{F} = \left\{ \sum_{i=1}^n \lambda_i f_i : \sum_{i=1}^n \lambda_i = 1, \lambda_i \geq 0, \ f_i \in \mathbb{F}, \ n \geq 1 \right\}.$$

Then, $\mathbb{F}$ can also be seen as the set of mixed strategies for $J_1$.

Now, for $J_2$, the decision model is of the form $(X, B, Q, R_2)$, where $X = \{\chi_0, \chi_1\}$, $0 < \chi_0 < \chi_1$, $B = B(\chi) = [0, 1]$, $\chi \in X$. The transition law $Q$ is given as in (24–27), and the reward is given by the function $R_2 = U_2$ with

$$U_2(b) = a\mu_2 + (1 - b)r_2 - \frac{k_2}{2}b^2\sigma_2^2,$$

where $b \in [0, 1]$, and it is also supposed that $0 < \mu_2 - r_2 < k_2\sigma_2^2$. Hence, $max_{b \in [0,1]}U_2(b)$ is attained for $b^* \in (0, 1)$ given by

$$b^* = \frac{\mu_2 - r_2}{k_2\sigma_2^2}.$$

Let $\mathbb{G}$ be the corresponding set of stationary (or mixed) strategies for $J_2$.

The game plays out as follows. Given an initial state $x_0 \in X$, both players take a decision $a_0 \in A(x_0)$ and $b_0 \in B(x_0)$ according to their mixed strategies $f$ and $g$. Then each player receives an expected reward $E_{x_0}^{f,g}[U_1(x_0, a_0, b_0)]$ and $E_{x_0}^{f,g}[U_2(x_0, a_0, b_0)]$, respectively. The game then changes to a new state $x_1 \in X$ according to the transition $Q(\cdot|x_0)$ and then the process repeats. In time, both players will receive the total of their expected rewards for each decision taken during the game, that is, they will receive

$$V_{J_1}(\chi, f, g) = \sum_{t=0}^\infty \eta^t E_{\chi,f,g}[U_1(f(x_t)] \text{ and } V_{J_2}(\chi, f, g) = \sum_{t=0}^\infty \beta^t E_{\chi,f,g}[U_2(g(x_t)],$$

respectively, where $x_0 = \chi$. Note that the game described constitutes a discounted stochastic game between two players in which they take decisions independently from each other and simultaneously.

Next, a pair of strategies $(f^*, g^*)$ is called a *Nash-equilibrium* if

$$V_{J_1}(\chi, f^*, g^*) = \sup_{f' \in \mathbb{F}} V_{J_1}(\chi, f', g^*)$$

and

$$V_{J_2}(\chi, f^*, g^*) = \sup_{g' \in \mathbb{G}} V_{J_2}(\chi, f^*, g'),$$

for each $\chi \in X$.

**Lemma 8.** *For the two-person game, the pair* $(f^*, g^*)$ *with* $f^*(\chi) = a^*$ *and* $g^*(\chi) = b^*$, *for all* $\chi \in X$ *is a Nash-equilibrium, and*

$$V_{J_1}(\chi, f^*, g^*) = \frac{U_1(a^*)}{1 - \beta}$$

*and*

$$V_{J_2}(\chi, f^*, g^*) = \frac{U_2(b^*)}{1 - \beta},$$

*for all* $\chi \in X$.

*Proof.* Observe that, under the conditions of Portfolio Choice problem, for each $f \in \mathbb{F}$, $g \in \mathbb{G}$, and $x_0 = \chi$,

$$V_{J_1}(\chi, f, g) = \sum_{t=0}^{\infty} \beta^t E_{\chi, f}[U_1(f(x_t)]$$

and

$$V_{J_2}(\chi, f, g) = \sum_{t=0}^{\infty} \beta^t E_{\chi, g}[U_2(g(x_t)].$$

Therefore, a direct application of Lemma 7 and Theorem 5 allow to obtain the proof of the following result.

**Lemma 9.** *Suppose that the fuzzy reward function is given by*

$$\tilde{R}(x, a) = (\gamma_1, \gamma_2, \gamma_3, \gamma_4) R(x, a),$$

*with* $(x, a) \in \mathbb{K}$. *Then, the fuzzy version of the Two-Person game, the Nash equilibrium is given by* $(a^*, b^*)$ *and* $\tilde{V}_{J_1}(\chi) = V_{J_1}(\chi, f^*, g^*)(\gamma_1, \gamma_2, \gamma_3, \gamma_4)$ *and* $\tilde{V}_{J_2}(\chi) = V_{J_2}(\chi, f^*, g^*)(\gamma_1, \gamma_2, \gamma_3, \gamma_4)$, *for all* $\chi \in X$.

## 6 Concluding Remark

Following ideas similar to those given in Sect. 4 of this paper, it is possible to obtain the optimal solution of the next optimal control problem. Consider the standard decision models given by

$$M_1 = (X, A, \{A(x) : x \in X\}, Q, R_1), \tag{29}$$

and

$$M_2 = (X, A, \{A(x) : x \in X\}, Q, R_2), \tag{30}$$

where both models satisfy the assumptions given in Sect. 3, and

$$0 < R_1(x, a) \le R_2(x, a) < \gamma, \tag{31}$$

for all $x \in X$, $a \in A(x)$, $\gamma$ is a positive constant, and $R_2 = zR_1$, $z > 1$

Now, take into account the infinite-horizon fuzzy optimal control problem with decision model:

$$\widetilde{M} = (X, A, \{A(x) : x \in X\}, Q, \widetilde{R}),\tag{32}$$

with

$$\widetilde{R}(x, a) := (0, R_1(x, a), R_2(x, a), \gamma),\tag{33}$$

$x \in X, a \in A(x)$. Notice that $\widetilde{R}$ given in (33) models the fact that, in fuzzy sense, "the reward is approximately in the interval $[R_1(x, a), R_2(x, a)], x \in X, a \in A(x)$".

Let $v_i$, $V_i$ and $f_i^*$ be the objective function, the optimal value function and the optimal stationary policy, respectively for the model $M_i, i = 1, 2$ and let $\widetilde{V}$ by the optimal value function for $\widetilde{M}$. As in the proof of Theorem 4, using that for each $\pi \in \mathbb{F}$ and $x \in X$,

$$v_1(\pi, x) \le v_2(\pi, x)\tag{34}$$

and it is direct to obtain that, for each $\pi \in \mathbb{F}$, $x \in X$ and $\alpha$ :

$$\alpha v_1(\pi, x) \le \alpha v_2(\pi, x) \le \alpha v_2(f_2^*, x) = \alpha V_2(x),\tag{35}$$

and

$$\alpha v_2(\pi, x) + (1 - \alpha)(\frac{\gamma}{1 - \beta}) \le \alpha v_2(f_2^*, x) + (1 - \alpha)(\frac{\gamma}{1 - \beta})$$

$$= \alpha V_2(x) + (1 - \alpha)(\frac{\gamma}{1 - \beta}).\tag{36}$$

Hence, from (35) and (36) it results that

$$\widetilde{V}(x) = \left(0, V_2(x), V_2(x), \frac{\gamma}{1 - \beta}\right),\tag{37}$$

$x \in X$, and $f_2^* = f_1^*$ is optimal for $\widetilde{M}$.

Observe that $\widetilde{V}$ can be seen as the triangular type:

$$\widetilde{V}(x) = \left(0, V_2(x), \frac{\gamma}{1 - \beta}\right),\tag{38}$$

$x \in X$.

# References

1. Abbasbandy, S., Hajjari, T.: A new approach for ranking of trapezoidal fuzzy numbers. Comput. Math. Appl. **57**(3), 413–419 (2009)
2. Aliprantis, C.D., Border, K.: Infinite Dimensional Analysis. Springer, Heidelberg (2006). https://doi.org/10.1007/3-540-29587-9.pdf
3. Asmussen, S.: Applied Probability and Queues. Wiley, New York (1987). https://doi.org/10.1007/b97236
4. Carrero-Vera, K., Cruz-Suárez, H., Montes-de-Oca, R.: Discounted Markov decision processes with fuzzy rewards induced by non-fuzzy systems. In: Parlier G.H., Liberatore F., Demange, M. (eds.) ICORES 2021, Proceedings of the 10th International Conference on Operations Research and Enterprise Systems, pp. 49–59. SCITEPRESS (2021)
5. De Silva, C.W.: Intelligent Control: Fuzzy Logic Applications. CRC Press, Boca Raton (2018)
6. Dombi, J., Jónás, T.: Ranking trapezoidal fuzzy numbers using a parametric relation pair. Fuzzy Sets Syst. **399**, 20–43 (2020)
7. Driankov, D., Hellendoorn, H., Reinfrank, M.: An Introduction to Fuzzy Control. Springer Science & Business Media, Heidelberg (2013). https://doi.org/10.1007/978-3-662-11131-4
8. Efendi, R., Arbaiy, N., Deris, M.M.: A new procedure in stock market forecasting based on fuzzy random auto-regression time series model. Inf. Sci. **441**, 113–132 (2018)
9. Fakoor, M., Kosari, A., Jafarzadeh, M.: Humanoid robot path planning with fuzzy Markov decision processes. J. Appl. Res. Technol. **14**(5), 300–310 (2016)
10. Filar, J., Vrieze, K.: Competitive Markov Decision Processes. Springer Science & Business Media, Heidelberg (2012). https://doi.org/10.1007/978-1-4612-4054-9
11. Furukawa, N.: Parametric orders on fuzzy numbers and their roles in fuzzy optimization problems. Optimization **40**(2), 171–192 (1997)
12. Gil, M.A., Colubi, A., Terán, P.: Random fuzzy sets: why, when, how. Boletín de Estadística e Investigación Opeativa **30**(1), 5–29 (2014)
13. Guo, Y., Jiao, L., Wang, S., Wang, S., Liu, F., Hua, W.: Fuzzy superpixels for polarimetric SAR images classification. IEEE Trans. Fuzzy Syst. **26**(5), 2846–2860 (2018)
14. Hernández-Lerma, O.: Adaptive Markov Control Processes. Springer Science & Business Media, Heidelberg (1989). https://doi.org/10.1007/978-1-4419-8714-3
15. Klir, G.J., Yuan, B.: Fuzzy Sets and Fuzzy Logic: Theory and Applications. Prentice Hall, Upper Saddle River (1995)
16. Kurano, M., Yasuda, M., Nakagami, J., Yoshida, Y.: Markov decision processes with fuzzy rewards. J. Nonlinear Convex Anal. **4**(1), 105–116 (2003)
17. Ljungqvist, L., Sargent, T.J.: Recursive Macroeconomic Theory. MIT Press, Massachusetts (2012)
18. Phuong, N.H., Kreinovich, V.: Fuzzy logic and its applications in medicine. Int. J. Med. Inf. **62**(2–3), 165–173 (2001)
19. Porteus, E.L.: Foundations of Stochastic Inventory Theory. Stanford Business Books, Stanford (2002)
20. Puri, M.L., Ralescu, D.A.: Fuzzy random variables. J. Math. Anal. Appl. **114**(2), 409–422 (1993)
21. Puterman, M.L.: Markov Decision Processes: Discrete Stochastic Dynamic Programming. Wiley, New Jersey (1994)
22. Ramík, J., Rimánek, J.: Inequality relation between fuzzy numbers and its use in fuzzy optimization. Fuzzy Sets Syst. **16**(2), 123–138 (1985)
23. Semmouri, A., Jourhmane, M., Belhallaj, Z.: Discounted Markov decision processes with fuzzy costs. Ann. Oper. Res. **295**(2), 769–786 (2020). https://doi.org/10.1007/s10479-020-03783-6

24. Shapley, L.S.: Stochastic games. Proc. Natl. Acad. Sci. **39**(10), 1095–1100 (1953)
25. Syropoulos, A., Grammenos, T.: A Modern Introduction to Fuzzy Mathematics. Wiley, New York (2020)
26. Topkis, D.M.: Supermodularity and Complementarity. Princeton University Press, New Jersey (1998)
27. Webb, J.N.: Game Theory: Decisions, Interaction and Evolution. Springer-Verlag, London (2007). https://doi.org/10.1007/978-1-84628-636-0
28. Zadeh, L.A.: Fuzzy sets. Inf. Control **8**(3), 338–353 (1965)

# The Optimal Stopping Criteria for a Customer Contact Strategy in Targeted Marketing

Shiva Ramoudith$^{(\boxtimes)}$ [iD], Patrick Hosein [iD], and Inzamam Rahaman [iD]

The University of the West Indies, St. Augustine, Trinidad
{shiva,inzamam}@lab.tt, patrick.hosein@sta.uwi.edu

**Abstract.** Many companies and institutions, such as banks, usually have a wide range of products which must be marketed to their customers. Multiple contact channels such as phone calls (the most common but also most costly), emails, postal mail and Social Media are used for marketing these products to specific customers. The more contacts (and hence cost to the company) made to a customer the higher the chance that the customer will subscribe but beyond a certain limit this customer may in fact become irritated by such calls if they are not really interested in the product (which is another potential cost to the company if they lose the customer). Previous work has shown that one can use historical data on customer contacts together with demographic information of those customers to significantly increase the average number of subscriptions achieved, or products bought, per phone call (or contact) made when considering new customers. We demonstrate an improved approach to this problem and illustrate with data obtained from a bank.

**Keywords:** Resource allocation · Marketing · Data analytics · Machine learning · Optimization

## 1 Introduction

Customers must be cognizant of a business' offerings if said business is to earn revenue. Consequently, marketing is an indispensable component of any contemporary business strategy, with there being evidence [9] that companies that maintain or increase their advertising expenditures through turbulent times, such as the current global pandemic, tend to fare better than those that downsize their advertising.

However, not all advertising tactics are created equal. Advertising can take many different forms such as email campaigns, social media ads, and phone calls. In addition, advertising can be tailored and targeted to different customers. At the end of the day, businesses care about their bottom line: how to advertise to create the greatest lift in revenue and profits. A central concept in much of advertising is the derivation of customer segments [12]. By dividing customers into segments, a business can better tailor advertising strategies and decide which customers they ought to actually target.

Telemarketing refers to marketing initiatives that involve directly contacting customers through phone calls [7]. Telemarketing is a useful tool as the business can receive rapid feedback on whether a customer is interested in a product or service. However,

© Springer Nature Switzerland AG 2022
G. H. Parlier et al. (Eds.): ICORES 2020/2021, CCIS 1623, pp. 193–210, 2022.
https://doi.org/10.1007/978-3-031-10725-2_10

customer segments can differ widely in their reactions to telemarketing campaigns [20]. As such, telemarketing campaigns would need to be precisely designed to omit certain segments from consideration [15] such that resources are not wasted and customers are not needlessly bothered [15]. Aside from the ethical concerns of wasting a customer's time, avoiding wastage of a customer's time is also pragmatic: an irritated customer is likely to resent the business, thereby forgoing the business' products and services. Given the above, it would be shrewd to design phone-based marketing campaigns to target specific customer segments. The design of such marketing campaigns ought to consider that the budget of contact attempts that can be made is limited.

In this paper, we formalize the problem of designing a customer segment-aware data-driven telemarketing campaign. Our formulation uses historical data to divide a customer base into segments and then allocates calls from a budget of calls to these different segments with the objective of maximizing the number of customers who buy into a product or service. In addition, our formulation considers how many times members of different segments ought to be contacted to avoid wasting resources and irritating customers. Our method was validated using the Portuguese bank dataset of Moro et al. [13]. Furthermore, it should be noted that this paper is an extension of Hosein et al. [5]. This paper is different from the original conference paper Hosein et al. [5] in that it proposes an improved method for deriving the customer segments; and further demonstrates the improvement in performance over the original paper in terms of success captured and run-time efficiency for the same number of calls.

## 2    Related Work and Contributions

Several papers have tackled the problem of designing telemarketing campaigns through the prism of operations research and data-science. Many of these papers also employ the telemarketing dataset curated by Moro et al. [13].

Most of the prior work formalize the problem as binary classification problem. In these formulations, authors endeavoured to determine whether or not a customer would accept the marketed product based on their features. Karim and Rahman [6] compared the use of the C4.5 decision tree algorithm [18] against the Naive Bayes algorithm and found that the former outperformed the latter. Similarly, Lawi et al. [10] compared SVMs against Ada-boosted SVMs. Moreover, they employed grid searches for their models; they found that their hyper-parameter tuned Ada-boosted SVM outperformed their hyper-parameter tuned SVM. Kozak et al. [8] devised an adaptation of the ant-colony decision forest algorithm and found that it performed well compared with non-ensemble baselines.

Along the lines of binary classification, some authors have also considered neural networks. Puteri et al. [17] compared the use of radial basis functions (RBF) as activation functions against Sigmoidal activation functions for determining customers to call in the framework of binary classification; they found that a network with RBF activations outperformed a network with Sigmoidal activations. Turkmen et al. [22] considered recurrent neural networks. More recently, Gu et al. [4] used multi-filter CNN to predict which customers would take up the offered product.

Neural Network models have also been examined and compared in some previous work. Puteri et al. [17] compared the use of radial basis functions (RBF) as activation functions against Sigmoidal activation functions for determining customers to call in the framework of binary classification. Puteri et al. [17] found that a network with RBF activations in the hidden layer performed better than Sigmoidal activations in the hidden layer.

Moro et al. [14] employed a data mining oriented approach that used sliding windows to compute measures of customer lifetime value (LTV). LTV is a proxy for a customer's value to the company based on the projected future interactions [2]. Using sensitivity analysis, they were able to derive explanations for which LTV measures were most important and demonstrated that LTV measures computed from historical data are useful in predicting future behaviour, thereby obviating the need for acquiring more information about customers.

Bertsimas and Mersereau [1] developed a dynamic programming formulation for allocating messages to multiple customer segments. In their paper, they also propose a Lagrangian relaxation of their initial dynamic programming problem and showed that their Lagrangian relaxation performs well in practice. They assumed that customer segments are known and so do not provide a procedure for the extraction of customer segments from data.

## 3   Mathematical Model

Suppose that we have set of customers along with the features of each customer (e.g. age, gender, occupation). In a telemarketing campaign, we can make several calls to each customer in an attempt to persuade them into purchasing a service or produce (e.g. loan). The customer might accept the product after a particular call, after which said customer is no longer considered a viable candidate for the duration of the campaign. Furthermore, after a number of calls, we remove a customer from the pool of potential customers.

In addition, suppose that the probability of a customer accepting the product can be expressed as function of the customer's features. Consequently, customers with similar features would have similar probabilities of accepting the product. Similar customers can be grouped together into customer segments. Firstly, we need to determine which customers should be targeted. Second, we need to determine how many times we should call a targeted customer before removing them from the pool of candidates.

Let us first consider a single customer segment before deciding how to allocate calls among segments. Suppose that we have $N$ customers in a segment. Moreover, let $p_i$ be the probability that a customer in this segment accepts on call $i$. If we can make a maximum of $k$ calls to a customer, let $s_k$ be the expected number of successes and $c_k$ be the the expected number of calls made. Note that Eqs. 1, 2, 3 and 4 are referenced from Hosein et al. [5]. From the above, the expected number of successes is given by

$$s_k = N \left( 1 - \prod_{i=1}^{k} (1 - p_i) \right) \tag{1}$$

and the expected number of calls is given by

$$c_k = N \left( k \prod_{j=1}^{k}(1 - p_j) + p_1 + \sum_{i=2}^{k} i p_i \prod_{j=1}^{i-1}(1 - p_j) \right) \tag{2}$$

Next consider the function $s_k$ versus $c_k$ as $k$ varies. The gradient of this function at some $k$ is given by

$$\frac{s_{k+1} - s_k}{c_{k+1} - c_k} = \frac{p_{k+1} \prod_{j=1}^{k}(1 - p_j)}{\prod_{j=1}^{k}(1 - p_j)} = p_{k+1} \tag{3}$$

Note that, in practice, $p_k$ decreases with increasing $k$ since a customer is less likely to purchase in successive attempts. Consider the piece-wise linear function with values at points $(c_k, s_k)$. The gradients of the successive linear components will be decreasing and hence this is a piece-wise linear concave function. Note that we assume here that, once we increase the maximum number of calls for one customer in this customer segment it is increased for all. This convexity property is important since it now implies that when we optimize over customer segments we will be solving a convex optimization problem. Please note as both $s_k$ and $c_k$ varies by $N$ then the plot does not change with $N$ but the axes are scaled accordingly.

The objective of the problem is to determine how many calls to assign to customers in each customer segment. Let $x_j$ denote the number of calls that are assigned to customer segment $j$. Let $S_j(x_j)$ denote the expected number of successes if $x_j$ calls are allocated to customer segment $j = 1 \ldots M$. This is the piece-wise linear, concave function that we derived above scaled by the number of members in $j$. The optimization problem becomes

$$V = \max_{\vec{x}} \sum_{j=1}^{M} S_j(x_j)$$

$$\text{s.t.} \sum_{j=1}^{M} x_j = T \tag{4}$$

$$\vec{x} \in \{0, \ldots, T\}^M$$

Since $S_j$ is piece-linear and concave one can show that a greedy approach can find a near-optimal solution to this problem. Find the customer segment with the largest initial gradient, assign as many calls as needed to get to the next break-point, update the gradient for that customer segment to that of the next linear segment and repeat until all $T$ calls have been allocated. If the last assignment brings the total calls for the customer segment to the next break-point, then the solution is optimal, and so the solution is typically quite close to being optimal. The pseudo-code provided in Fig. 1 can be used to determine the optimal allocation.

Let us illustrate with a simple example. Assume that we have two segments, $j = 1, 2$. For the first customer segment, suppose that the probability of success for each of the first five call attempts is 0.10, 0.07, 0.02, 0.01 and 0, respectively. Hence 20% of customers eventually accept, and the rest reject the offer. For the second customer

```
Require: T = Total number of calls to be allocated
Require: M = Total number of customer segments
Require: s_j(k) = Expected #successes for a maximum of k calls
Require: c_j(k) = Expected #calls for a maximum of k calls
Require: N_j = Number of customers in customer segment j
Require: k_j = 0 Initial maximum call value for customer segment j
Require: g_j(k_j)    =    s_j(1)/c_j(1) set initial gradient for customer
    segment j
  while T > 0 do
    j* = arg max_j{g_j(k_j)}
    k_j* ← k_j* + 1
    g_j* = (s_j*(k_j*+1)-s_j*(k_j*))/(c_j*(k_j*+1)-c_j*(k_j*))
    T = T - N_j*
  end while
  for j = 1 : M do
    x_j = N_j k_j
  end for
  return  x⃗
```

Fig. 1. Call allocation optimization pseudo-code [5].

segment, we assume that the corresponding probabilities are 0.06, 0.03, 0.02,0.01 and 0. The success versus call attempt plot is shown in Fig. 2.

Now suppose that we have 500 new customers with 200 being in customer segment 1 (i.e., $N = 200$ for this customer segment) and 300 in customer segment 2. The initial gradients are 0.10 and 0.06, respectively and hence the first set of customers are chosen from customer segment 1. 200, one for each customer in the customer segment are necessary to get to the first break-point. The gradient for customer segment 1 drops to 0.07, so it is chosen again, but this time 180 calls are needed to get to the next break-point. At this break-point, the gradient drops to 0.02, so customer segment 2 is chosen next and 300 calls are required to get to the first break-point. Hence, if we assumed a budget of 680 calls, then it is optimal to call every customer in segment 1 no more than twice and to call every customer in segment 2 once.

## 4 Numerical Results

In this section we provide numerical results to illustrate improvement over the existing approach by Hosein et al. [5] in a simulated environment. We firstly describe the transformations applied to the dataset and then provide details on the implementation of the improved approach. We finally compare the methodologies in terms of area under the curve (AUC) and run-time performance.

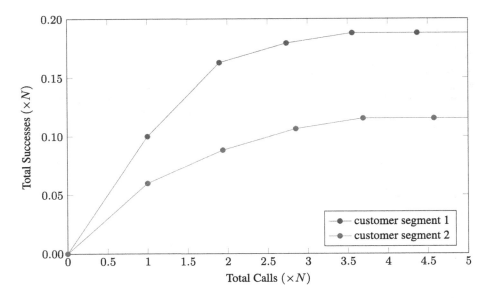

**Fig. 2.** Variation of successes with calls [5].

## 4.1  Data Description

We used the Bank Marketing Dataset collected by Moro et al. [13]. The dataset is available on the UCI Machine Learning Repository. The dataset contains customers from a Portuguese banking institution and the direct marketing campaigns used to encourage them to subscribe to a term-deposit product. The dataset contains 45211 records, each representing a customer contacted during the marketing campaign.

Table 1 presents the client level features present within the dataset. In instances where data was missing for certain features, they were set to Unknown.

Additional features include data concerning the last contact within the current campaign, the previous campaign outcome, and social and economic context attributes. Features such as the contact medium and previous campaign outcome contain over 25% and 81% null values, respectively, resulting in these and relating features being omitted. We did not include the social and economic context attributes for clustering since they do not represent client-level data. We filter potentially anomalous data points from our analysis by considering 99% of all calls made (we ignore customers contacted greater than 34 times).

Figure 3 contains a plot of the successes divided by the number of calls for each call number. We notice that increases in the frequency of contact result in a decreased likelihood of a customer subscribing to a product or service. In some instances, repeated customer contacts can irritate them, resulting in increased unwillingness in other campaigns or potentially the customer closing their bank account.

Our objective is to determine the order in which customers should be called and how often they should be called. We utilize the data present to derive the customer segments and then determine how many calls to allocate to each customer within a customer

**Table 1.** Customer features with abbreviations for values of categorical features.

| Attribute | Values |
|---|---|
| Age | Customer's age |
| Balance | Customer's average yearly balance (€) |
| Job | Administrator (A), Blue-collar (BC), Entrepreneur (E), Housemaid (H), Management (M), Retired (R), Self-employed (SE), Services (S), Student (ST), Technician (T), Unemployed (UE) or Unknown (U) |
| Marital status | Married (M), Single (S), Divorced (D) or Unknown (U) |
| Education | Primary (P), Secondary (S), Tertiary (T) |
| Risk | Yes (Y), No (N) or Unknown (U) |
| Housing | Yes (Y), No (N) or Unknown (U) |
| Personal | Yes (Y), No (N) or Unknown (U) |
| Number of calls | 1–63 |
| Success of offer | Yes or No |

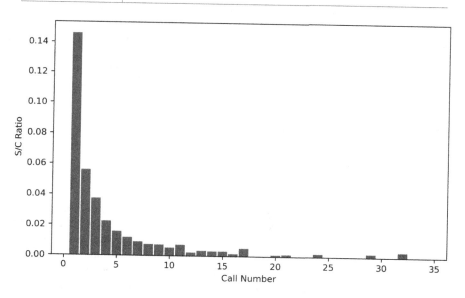

**Fig. 3.** Success ratio per call number.

segment. We firstly describe the transformations applied to the data, then the derivation of the customer segments. Finally, we demonstrate the performance with our improved approach through a simulation where we assign calls to customers.

## 4.2 Dataset Preprocessing

We selected all features mentioned in Table 1 for our analysis. The difference between our modified approach and the approach in [5] is the formation of customer segments.

We utilised a clustering algorithm instead of generating customer segments based on all combinations of values across features.

Before we could apply the clustering algorithm, we applied a series of transformations to the dataset. We first applied one-hot encoding to all categorical features since machine learning algorithms cannot directly work with categorical data. Moreover, we selected one hot encoding since these features are not ordinal. Secondly, we normalised the values for the continuous features since they have significant variations in their ranges. These features were age (18 to 95) and balance (−8019.0 to 102127.0). We applied this using the MinMaxScaler function from the Scikit-learn library [16]. We also sought to reduce the dimensionality of the data using Principal Component Analysis (PCA) [3] in hopes of uncovering additional information concerning the relationships between dependent variables. PCA accomplishes this by identifying directions, called principal components, which account for the maximum variability in the data as much as possible.

### 4.3   Customer Segment Assignment

We can use a customer's features to assign them to the appropriate customer segment or cluster. Before we can do this, we need to have a predefined set of customer segments. As aforementioned, we proposed the use of a clustering algorithm to form the customer segments. The selected algorithm is K-means [11]. The K-means algorithm tries to minimize the differences within a cluster while maximizing the distance among clusters. Our reasons for selecting K-means are as follows: it scales to large datasets efficiently, convergence is guaranteed, and it is relatively simple to implement. We are also aware that we must select the value for the hyper-parameter corresponding to the number of clusters before running the algorithm and issues concerning outliers and the increasing number of dimensions. In summary, it is more robust than the approach used to form the customer segments in Hosein et al. [5]. The referenced approach can only assign customers to clusters based on the exact match of a customer's features to a customer segment.

We now discuss how we selected the optimal number of clusters for our approach. We ran multiple iterations of the K-means algorithm with default parameters while adjusting the number of clusters, $2 <= k <= 600$. For each run, we utilized the exact subsets from the dataset for training and testing. We used the majority of the data to train the K-means algorithm. Mapping each data point from the training set to its corresponding cluster allowed us to derive the following information for each cluster:

- the number of successes captured for each call number ($s_i$)
- the number of calls made for each call number ($c_i$)
- the success per call ratio for each call number ($s_i/c_i$)
- the convex hull of $s_i$ vs $c_i$

We used the testing data to conduct simulations of contacting customers based on the customer segments the algorithm assigned each data point in our test set. We prioritized the contacts based on the overall $s/c$ ratio for each cluster. To determine the effectiveness from varying $K$, we computed the AUC for each value of $K$ from the successes and corresponding calls generated during each simulation. We set the optimal cluster number to the value of $K$ corresponding to the highest AUC value. Note that we applied PCA with varying values for the parameter *n_components* to our dataset before being input into the K-means algorithm in hopes of ignoring irrelevant features, which could potentially lead to increased performance.

When applying PCA, we sought to keep high explained variance ratios $\geq 0.85$. We attempted experiments using PCA on the data with ratios of 0.85 and 0.95. We also experimented without the use of PCA. Figure 4 shows the explained variance ratios and the associated number of components.

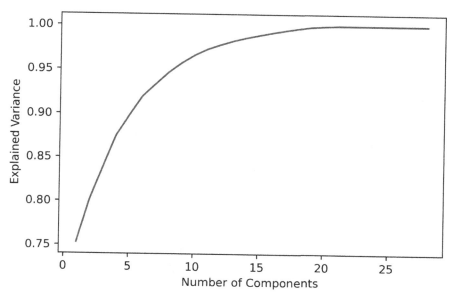

**Fig. 4.** Explained variance ratio for various numbers of components.

Referring to Fig. 5, we notice that the average AUC tends to stabilize around cluster number 100 regardless of the type of transformation applied to the data. After analyzing the results, the optimal values for $K$ were 175 (No PCA), 287 (PCA - 0.95) and 121 (PCA - 0.85). The AUC of these scores were 20611291.1, 20706138.2 and 20702395.6 units, respectively, with $\sigma$ of 53712.2 units, indicating that applying PCA did not significantly impact results. As a result, we chose not to use PCA and decided upon the optimal value for $K$ being 175. We also investigated the performance when selecting smaller cluster numbers within one $\sigma$ of the mean. We hypothesized that setting $k$ to a lower number would result in each cluster having a higher number of training data points which may increase performance when applied with the 'Gradient Ascent Approach'. We describe these experiments further on.

## 4.4    Optimal Allocation of Calls

We use the features present for each customer $i$ and the historical data on the number of attempts made to them and their outcome (success or failure) to compute relevant infor-

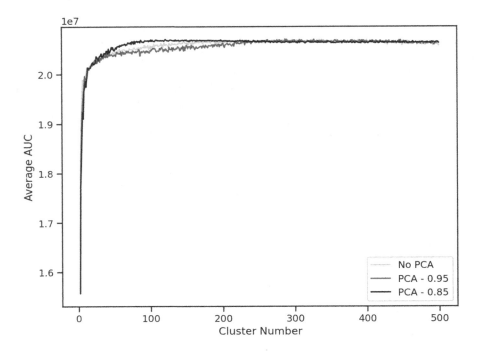

**Fig. 5.** Average AUC for each cluster number.

mation concerning each customer segment. Let $v_i$ denote the number of call attempts made to customer $i$. Let $q_i = 1$ if attempts to customer $i$ are successful (i.e., success was achieved on call $v_i$) and $q_i = 0$ otherwise. We assume $M$ customer segments, and these are indexed by $j$. We use $G_j$ to represent the set of users who have the features specified by customer segment $j$.

In order to reduce the number of calls made, one can reduce the maximum number of calls allocated to each customer. There will be a reduction in the number of successes captured, potentially resulting in a higher success per call rate. Let us introduce the variable $k$ as the specified maximum number of calls. If we reduce $k$, the institution might be unsuccessful in capturing success for certain customers. Equations 5 and 6 are referenced from Hosein et al. [5]. The number of successes achieved for a call maximum of $k$ for customer segment $j$ is given by:

$$s_k(j) = \sum_{i \in G_j} q_i[\min\{1, \max\{0, k - v_i + 1\}\}] \tag{5}$$

and the number of call attempts made is given by

$$c_k(j) = \sum_{i \in G_j} \min\{k, v_i\}] \tag{6}$$

Here we describe the methodology for determining the performance with our proposed approach. As previously mentioned, the simulation process involves contacting customers assigned to the customer segments in descending order of their overall $s/c$ ratio. This method prioritises the contacts made to customer segments (we initiate contact with those who are more likely to subscribe to the term deposit first). It is the first variation and is termed the 'Greedy Approach (GC)' by Hosein et al. [5]. The alternative variation is to optimise within each customer segment, and this is termed the 'Gradient Ascent Approach (GA)' by Hosein et al. [5]. This variation involves computing the $s_i/c_i$ ratio at each successive call number, $i$, and using these values to prioritise calls. We can now assign calls to multiple customer segments at different call points instead of assigning calls to an entire customer segment at once. The latter variation is more computationally expensive but proposes a higher overall AUC score.

## 4.5 Experiment Setup

As previously mentioned, we utilized the K-means algorithm to derive the customer segments. Executing the experiments consumed a fair amount of time using the CPU based library Scikit-learn [16]. We opted to use RAPIDS [21], a recently developed machine learning library that facilitates the execution of end to end data science pipelines entirely on the GPU. We saw approximately 14x reduction in experiment runtime using RAPIDS (605 min for vs 43 min for $2 <= K <= 500$) for resolving the optimal number of customer segments. We conducted all experiments using an Intel i9-9900k CPU(turbo boost disabled), 64GB memory and, a Titan RTX GPU.

In order to obtain the average AUC for each cluster number, we utilized stratified 5-fold cross-validation on the dataset. We used 80% of the data from each split to train the clustering algorithm and construct customer segment metrics. We used the remaining 20% of the data to simulate calls to customers. We ensured that identical data points were used for training and testing by specifying the same *random_state* for the cross-validation process on the dataset. In addition, we also specified the same *random_state* for the K-means algorithm. Our reason for this is to aid in experiment reproducibility and serve as a base to compare the various experiments. To demonstrate the robustness of the approaches, we also varied the *random_state* for both cross-validation and K-means in further experiments.

The GA* approach requires that the gradients at each successive call number be decreasing or concave. There may be instances where the gradients for specific customer segments might fluctuate, preventing us from executing the GA* approach. As a result, we computed the convex hull for each customer segment. We utilized this function from the SciPy library [23]. If the convex hull failed to generate, we defaulted to using the overall $s/c$ ratio for that particular customer segment and contacted all customers.

To demonstrate the improvement with the proposed approach, we conducted six experiments. Firstly, for a Baseline and an Upper Bound on achievable performance, we re-ran these approaches provided by Hosein et al. [5]. We briefly define the Baseline and Upper Bound for the problem, denoted by BL and UB, respectively. BL represents the performance by simply calling all persons in the test set randomly. In contrast, UB represents the maximum performance possible; i.e. if we were to know how many times we should contact a customer and whether they would accept the product. Afterwards, we re-ran the Greedy and Gradient Ascent approaches by Hosein et al. [5]. Following this, we executed our variants to these approaches, termed GC* and GA*, respectively, with our proposed clustering approach. All these approaches return the number of successes and corresponding calls made for customers in the testing set.

### 4.6 Results and Discussion

Here we present the results for two of the five cross-validation folds. The results were similar across the remaining folds for all methods. As mentioned previously, we used the same data points for training and testing for each of the six methods. In order to compare the performance among the methods, we considered the resulting AUC of all approaches in relation to the AUC of the BL approach. Note that GC* and GA* were both executed with $K$ set to 175.

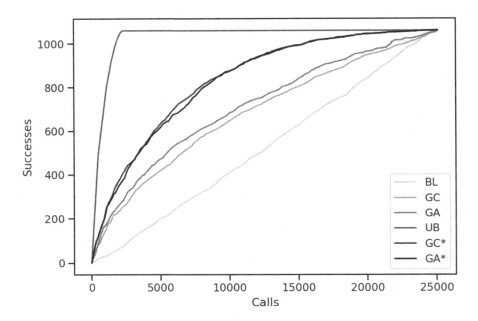

**Fig. 6.** Fold 1 - successes vs calls for multiple experiments.

We notice a significant increase in performance with GC* and GA* compared to GC and GA. We consider GA only since it represents the best performance possible with the alternative approach. GC* and GA* can obtain an average improvement in successes of 57.4% and 56.8% respectively over the Baseline approach, while GA can only achieve an average of 35.4% improvement. Note that the number of calls made is the same for all approaches. Selecting the K-means algorithm to build customer segments yields higher performance.

We expected GA* to be more effective than GC* however, GA* was more effective on average by 0.6%. This unexpected reduction in performance results from the convex

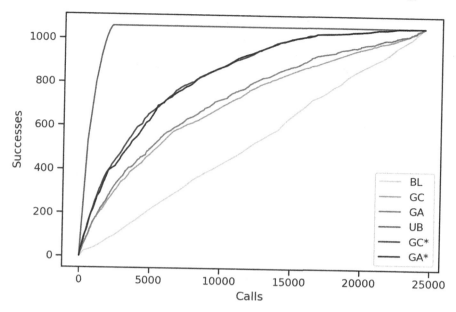

**Fig. 7.** Fold 3 - successes vs calls for multiple experiments.

hull not being constructed for several customer segments in GA*. Of the 175 customer segments, there were, on average, 74 customer segments across the five cross-validation folds where the convex hull failed to execute. The reason for this is that these clusters had very little change in the number of successes and calls for each call number, effectively having less than 3 data points for input into the convex hull. In this case, we resorted to using the overall $s/c$ ratio when computing the gradients. Moreover, these segments contained on average 66.2% of the successes from the testing set across the five cross-validation folds, leaving a mere third of the testing data where the gradient-based approach could be applied.

We then attempted further runs of the GA* approach for a series of cluster numbers that correspond to the AUC within a range of one $\sigma$ of the mean AUC for cluster number 175. These cluster numbers were smaller (16, 26, 42 and 84) to reduce the number of customer segments where the convex hull failed to generate. In summary, this did reduce the number of failed instances on average by 10% with cluster number set to 16. However, this did not positively affect the performance. The K-means algorithm effectively bundled the customer segments where the convex hull failed to generate into fewer ones. The average AUC for each selected cluster number mentioned was also lower than that of cluster number 175. We notice that the performance when using cluster number 84 was very close to that of cluster number 175, having 99.2% of its AUC. Adjusting the number of customer segments could not easily be applied with the approach by Hosein et al. [5] as one has to manually adjust the groupings of values within a feature to reduce the number of customer segments generated. This property of our proposed approach is helpful as the business (depending on the use case) can make customer segmentation more/less granular.

Lastly, to demonstrate robustness, we executed 25 GA* and GC* iterations with random seeds for both the cross-validation and K-means clustering. For GC*, we had a mean of 20533619.4 and corresponding $\sigma$ being 24012.2 units. With GA*, we had a mean of 20428402.6 and corresponding $\sigma$ being 25525.9 units.

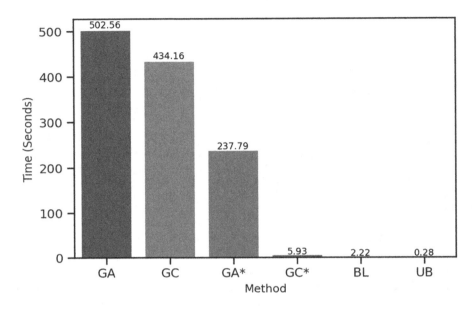

**Fig. 8.** Experiment run-times for each method over 5 cross-validation folds.

According to Fig. 8, we notice a significant decrease in run-time for our improved approaches. Regarding time complexity, the approach used by Hosein et al. [5] for generating the possible customer segments is exponential; this involves computing all possible combinations of unique attributes across all features. We provide an example to illustrate this. Let us assume we have ten customers with two features, education, denoted by $f_1$ (can be split into primary, secondary and tertiary), and married, denoted by $f_2$ (can be split into yes and no), we can construct the possible segments by all combinations of unique values across each feature. We represent the number of unique values for a feature by the function $v(f)$. The number of unique values in features $v(f_1)$ and $v(f_2)$ are 3 and 2, respectively. Therefore, we have a total of $v(f_1) * v(f_2)$ possible segments. In general, given a dataset, with features ranging from 1 to $m$, we would have the total number of customer segments, $cs_{old}$, being:

$$cs_{old} = \prod_{i=1}^{m} v(f_i) \tag{7}$$

With this dataset, the approach by Hosein et al. [5] had 1152 customer segments, with over 65% of them not having any customers assigned. There would be more unique values per feature given a larger dataset, resulting in an explosion of customer segments. The authors tried to mitigate this by grouping specific values within a continuous feature as one value through clustering algorithms; however, this is not always guaranteed to work. In terms of our approach, we are limited to the time complexity of the K-means algorithm, with average complexity for experimenting with one cluster number given by $O(KnT)$ where $K$ is the number of clusters, $n$ is the number of samples, and $T$ is the number of iterations. Our approach in most circumstances is less computationally expensive than the approach by Hosein et al. [5]. One caveat with our approach is that we have to determine the optimal cluster number by conducting repetitive experiments for each cluster number and imposing a maximum limit, $M$, on the number of clusters in our search space. As a result, the time complexity for our approach is $O(MknT)$. The optimal number of customer segments, $cs_{new}$, for the GC* approach is denoted by:

$$cs_{new} = \underset{k \in \{2 \leq k \leq M | k \in \mathbb{Z}\}}{\arg\max} \ (\text{AUC}(\text{GC}^*(k))) \tag{8}$$

The remaining logic for iterating the customer segments and simulating calls to customers for both approaches remain the same and is bounded by $cs_{old}$ and $cs_{new}$ for the respective approaches. Lastly, both approaches have similar storage requirements for both the greedy and gradient ascent approaches.

Table 2 provides a breakdown of customer segments, randomly chosen, from best to worst, for the first cross-validation fold using our proposed approach of segmenting the customer base. For the categorical features, we only list the attributes that comprised the majority of the feature, along with their percentage. Table 1 provides a breakdown of the abbreviations for the values concerning each feature. We discuss trends observed across the range of customer segments, ordered by the $s/c$ ratio.

We notice that customer segments having higher $s/c$ ratios are associated with customers with higher account balances and higher levels of education (tertiary and secondary) compared to primary education. Higher education qualifications translate into our observations regarding the types of jobs these customers have (most based in management, technician, and in some cases blue collar areas). We also see clusters with high $s/c$ ratios containing students and retired persons entirely, indicating that both young and elderly persons are interested in investing in the term deposit product. Furthermore, we notice that the high $s/c$ ratio customer segments do not have any existing loans; very few, however, have defaulted on loans. As we go to the lower-ranked customer segments, we notice an increase in customers having housing loans and the customer defaulting on a previous term deposit product and low account balances, which indicate that these customers do not possess the financial resources to subscribe to the term deposit product.

Half of the top 10 customer segments include predominantly married customers; in some cases, married persons may reside in households with two income sources, so these persons may be more inclined to purchase the term deposit product. Out of the top 10 customer segments, customer segment 3 had the most deviation from the rest. It consisted only of retired persons who were divorced and had high account balances; these types of customers are rare, and according to their marital status, we infer they would have less expense than other persons in the same age grouping, making them more likely to subscribe to the term deposit product.

When compared to the existing customer segmentation approach by Hosein et al. [5], we notice that both methods detected a customer segment containing mainly retired persons. This was the highest-ranked customer segment in the approach by Hosein et al. [5], having a $s/c$ ratio of 1.0. Further investigation reveals that this customer segment contained only two calls and two successes. In contrast, our approach ranked a similar customer segment as third, having a $s/c$ ratio of 0.312 and contained 215 calls and 67 successes. Despite our segmentation approach having a smaller $s/c$ ratio, we have more data points assigned to our customer segment, making inference more reputable. Furthermore, we achieve better performance overall, indicating that our customer segmentation approach is more effective. We also investigate customer segments at a deeper level since there is no grouping of values within a feature, as we saw with the alternative approach.

**Table 2.** Some sample customer segments for one of the cross validation folds (ranked by success rate). abbreviations for the attributes are defined in Table 1.

| CS | Rate | Avg. Age | Avg. Bal. | Edu. | Marital status | Job | Risk | Personal | Housing |
|---|---|---|---|---|---|---|---|---|---|
| 1 | 0.334 | 33 | 1990.81 | T - 100% | S - 100% | M - 45.6% T - 14% SE - 10.1% | N - 99.7% | N - 100% | N - 100% |
| 3 | 0.312 | 67 | 1931.74 | T - 100% | D - 100% | R - 100% | N - 100% | N - 100% | N - 100% |
| 20 | 0.096 | 33 | 1754.62 | T - 100% | M - 100% | T - 50.6% SE - 10.5% ST - 10.1% | N - 99.2% | N - 100% | N - 100% |
| 60 | 0.039 | 34 | 1651.47 | T - 97.9% | S - 100% | M - 71.8% SE - 7.3% E - 6% | N - 98.3% | Y - 100% | N - 100% |
| 100 | 0.017 | 47 | 1294.00 | P - 50.1% | D - 100% | BC - 28.7% R - 15.7% T - 15.7% | N - 96.3% | Y - 100% | N - 94.4% |
| 140 | 0.004 | 39 | 981.09 | S - 100% | S - 69.6% | T - 21.7% A - 21.7% S - 17.4% | N - 100% | N - 100% | N - 91.3% |
| 150 | 0.002 | 43 | 858.13 | S - 82.89% | D - 100% | A - 23.7% T - 21.1% S - 17.1% | N - 97.4% | N - 100% | Y - 100% |

### 4.7  Deployment

The code base with our implementation is publicly available at the GitHub repository [19]. Once deployed, we can apply our method periodically (e.g. once per week) to determine suitable customers to contact given a limited call budget. Agents would attempt contact with the selected customers based on their allowed number of calls.

After an extended period of time (e.g. two months), one can include new customer contact data into the dataset followed by a rebuild of the customer segments to ensure that the representation of the customer base contains the most up to date information. As more customer outcomes are collected, one can expect to improve the accuracy of the estimated probability distribution of each customer segment, thus enhancing decision making and further increasing campaign success.

## 5 Conclusion

Our results indicate that our improved method of segmenting customers using the K-means algorithm does improve the success of telemarketing campaigns given a limited budget of calls. In simulations, we noted increased performance compared to an existing approach by capturing 22% more success for the same number of calls. In addition, our proposed method offers reduced run-time and requires fewer hyperparameters, making it more robust and efficient.

## References

1. Bertsimas, D., Mersereau, A.J.: A learning approach for interactive marketing to a customer segment. Oper. Res. **55**(6), 1120–1135 (2007)
2. Dwyer, F.R.: Customer lifetime valuation to support marketing decision making. J. Direct Mark. **11**(4), 6–13 (1997)
3. Pearson, K.: Liii. on lines and planes of closest fit to systems of points in space. Lond. Edinb. Dublin Philos. Mag. J. Sci. **2**(11), 559–572 (1901). https://doi.org/10.1080/14786440109462720
4. Gu, J., Na, J., Park, J., Kim, H.: Predicting success of outbound telemarketing in insurance policy loans using an explainable multiple-filter convolutional neural network. Appl. Sci. **11**(15), 7147 (2021)
5. Hosein, P., Ramoudith, S., Rahaman, I.: On the optimal allocation of resources for a marketing campaign. In: ICORES, pp. 169–176 (2021)
6. Karim, M., Rahman, R.M.: Decision tree and Naive Bayes algorithm for classification and generation of actionable knowledge for direct marketing. J. Softw. Eng. Appl. **6**(4), 196–206 (2013)
7. Kotler, P., Keller, K.: A Framework for Marketing Management, 5th edn. Prentice Hall, Upper Saddle River (2011)
8. Kozak, J., Juszczuk, P.: The ACDF algorithm in the stream data analysis for the bank telemarketing campaign. In: 2018 5th International Conference on Soft Computing & Machine Intelligence (ISCMI), pp. 49–53. IEEE (2018)
9. Kumar, N., Pauwels, K.: Don't cut your marketing budget in a recession. Harvard Business Review (2020)
10. Lawi, A., Velayaty, A.A., Zainuddin, Z.: On identifying potential direct marketing consumers using adaptive boosted support vector machine. In: 2017 4th International Conference on Computer Applications and Information Processing Technology (CAIPT), pp. 1–4. IEEE, Kuta Bali, Indonesia (2017)
11. Lloyd, S.: Least squares quantization in PCM. IEEE Trans. Inf. Theory **28**(2), 129–137 (1982)
12. Loshin, D., Reifer, A.: Using Information to Develop a Culture of Customer Centricity: Customer Centricity, Analytics, and Information Utilization. Morgan Kaufmann Publishers Inc., San Francisco (2013)
13. Moro, S., Cortez, P., Rita, P.: A data-driven approach to predict the success of bank telemarketing. Decis. Support Syst. **62**, 22–31 (2014)

14. Moro, S., Cortez, P., Rita, P.: Using customer lifetime value and neural networks to improve the prediction of bank deposit subscription in telemarketing campaigns. Neural Comput. Appl. **26**(1), 131–139 (2014). https://doi.org/10.1007/s00521-014-1703-0
15. Mylonakis, J.: The influence of banking advertising on bank customers: an examination of Greek bank customers' choices. Banks Bank Syst. **3**(4), 44–49 (2008)
16. Pedregosa, F., et al.: Scikit-learn: machine learning in python. J. Mach. Learn. Res. **12**(10), 2825–2830 (2011)
17. Puteri, A.N., Tahir, Z., et al.: Comparison of potential telemarketing customers predictions with a data mining approach using the MLPNN and RBFNN methods. In: 2019 International Conference on Information and Communications Technology (ICOIACT), pp. 383–387. IEEE, IEEE, Yogyakarta, Indonesia (2019)
18. Quinlan, J.R.: C4. 5: Programs for Machine Learning. Elsevier, Amsterdam (2014)
19. Ramoudith, S., Rahaman, I., Hosein, P.: Implementation of improved customer segmentation approach. GitHub (2021). https://github.com/shiv1994/BankMarketingExtended
20. Roach, G.: Consumer perceptions of mobile phone marketing: a direct marketing innovation. Direct Market. Int. J. **3**(2), 124–138 (2009)
21. Team, R.D.: RAPIDS: collection of libraries for end to end GPU data science (2018). https://rapids.ai
22. Turkmen, E.: Deep learning based methods for processing data in telemarketing-success prediction. In: 2021 Third International Conference on Intelligent Communication Technologies and Virtual Mobile Networks (ICICV), pp. 1161–1166. IEEE (2021)
23. Virtanen, P., et al.: Scipy 10: fundamental algorithms for scientific computing in python. Nat. Methods **17**(3), 261–272 (2020)

# A MIP-Based Heuristic for Pickup and Delivery on Rectilinear Layout

Claudio Arbib[1]($\boxtimes$)(iD), Andrea Pizzuti[2](iD), Fatemeh K. Ranjbar[1](iD),
and Stefano Smriglio[1](iD)

[1] DISIM, Università degli Studi dell'Aquila, L'aquila, Italy
{claudio.arbib,stefano.smriglio}@univaq.it,
fatemeh.kafashranjbar@graduate.univaq.it
[2] DII, Università Politecnica delle Marche, Ancona, Italy
a.pizzuti@univpm.it

**Abstract.** The Vehicle Routing Problem with Pickups and Deliveries (VRPPD) arises in many application contexts and has been intensively studied in the last decades. We investigate the special case where pickup and delivery locations are distributed on a line. Although this situation is frequent when handling material in manufacturing systems with rectilinear layout, this case has not received enough attention so far. Derived from a real application, our general model also features load/unload times, vehicle capacities and the absence of a depot. A two-stage MIP-based heuristic that exploits such a special topology is devised, and its performance is assessed within an industrial case study provided by a large semiconductor manufacturer. We first compare our method to a standard Clarke and Wright type heuristic, then document its practical impact when implemented in a dynamic environment.

**Keywords:** Vehicle routing · Pickup and delivery · Mixed-integer programming

## 1 Introduction

We are given a set $I$ of $n$ locations placed on a line at fixed positions and a set $R$ of requests, each of which, say $r \in R$, asks for handling an amount $l_r$ of commodity from a *pickup location* $p_r \in I$ to a *delivery location* $d_r \in I$ (without intermediate transshipments at other locations). Handling is carried out by a set $K$ of $m$ vehicles. Each vehicle $k \in K$ travels at constant speed $v$ and has a limited loading capacity $c_k$. Finally, for a request $r \in R$, pickup/delivery operations take $u \cdot l_r$ time, where $u$ is a given constant. We wish to determine vehicle routes to process all the requests while minimizing the *makespan*, that is, the largest completion time among all requests.

This problem is a special case of the classical VEHICLE ROUTING PROBLEM WITH PICKUPS AND DELIVERIES (VRPPD), intensively studied in the last decades (for transportation services dedicated to people, one speaks of a DIAL-A-RIDE PROBLEM). We refer the reader to [2, 12] for an introduction to this huge body of research.

A distinguishing feature of our problem is the rectilinear layout of the pickup/delivery locations: so we will call it the RECTILINEAR-VRPPD (R-VRPPD). Such a

© Springer Nature Switzerland AG 2022
G. H. Parlier et al. (Eds.): ICORES 2020/2021, CCIS 1623, pp. 211–226, 2022.
https://doi.org/10.1007/978-3-031-10725-2_11

situation arises, for instance, in manufacturing plants where machining facilities are deployed along an aisle. This case has not received enough attention in the VRP literature; yet, the rectilinear layout has been investigated in related studies. [8] considers the problem of transporting a set of items, each with a specific origin and destination, along a path; the Author observes that minimizing the tour length is NP-hard if items cannot be dropped at intermediate stations, while it can be solved in linear time otherwise. [13] propose polynomial-time algorithms to solve a single-commodity pick-up and delivery problem on a path using a single vehicle with unit, or general finite, or unlimited capacity. Both studies do not capture the complexity of our case, where a fleet of vehicles is given, each requests involves a separate commodity, and items cannot be dropped at intermediate stations.

Another feature of our model is the way load/unload is considered. Such operation has in fact a double effect on solution, in terms of both capacity and time usage. While vehicle capacity is addressed in various papers (one for all, [11]), load/unload time is generally dealt with by incorporating it into the total time required to fulfill a request. However, an economy of scale can be gained by suitably grouping requests that share the origin and/or the destination (provided that enough capacity is available). To the best of our knowledge, both issues are not addressed simultaneously in the literature.

A third feature of our model is that it is designed for use in a dynamical environment, where the initial position of each vehicle can be arbitrary in the aisle. In fact, in the industrial case which stimulated this study [1], the production process is continuous and no location can a-priori be identified as the vehicle depot. Typical solution methods for the VRPPD are instead developed under the assumption of a single depot, from which vehicles start at the beginning of the work shift and where they go back at its end. This not irrelevant detail makes, for instance, standard *cluster-first route-second* heuristics not directly usable, as the cost of assigning a cluster to a vehicle is not independent on the vehicle but instead depends on its initial position.

Summarizing, our contribution is twofold: on one hand we focus on practical aspects (topology, load/unload operations, absence of a depot) not satisfactorily addressed in the VRP literature; in so doing, on the other hand, we propose a way to exploit such peculiarities and gain in both algorithm efficiency and solution accuracy. In particular, the rectilinear layout allows us to devise a math-heuristic that tackles the problem in two stages, according to the following *cluster-first route-second* strategy:

1. **Request Assignment (Clustering).** Via a MIXED INTEGER (LINEAR) PROGRAM (MIP), assign the requests in $R$ to as many distinct segments of the line, as available vehicles. Each segment will identify a route to be next assigned to some vehicle.

2. **Vehicle Assignment and Scheduling (Routing).** Schedule each vehicle on each distinct route and determine in this way the time required of the vehicle to fulfill all the requests assigned to that route. Then, by solving a BOTTLENECK MATCHING PROBLEM, find a one-to-one vehicle-to-route assignment that minimizes the largest among all vehicles completion time.

The whole heuristic is illustrated in Sect. 2. Its performance is investigated in Sect. 3 by an industrial case study provided by a large semiconductor manufacturer, first illustrated in [1], which motivated the present study. Here we address the RVRPPD problem along with the proposed decomposition framework from a more general perspective. In particular, we document a comparison of our heuristic to a standard Clarke and Wright's *cluster-first route-second* heuristic for the VRP (adequately modified to match the problem features inherited from the application). This is done in a *static* environment, where all requests are known in advance. Then, we review the adaptation of the heuristic to a dynamic context, as in [1]: now the problem becomes an R-VRPPD with time windows, as requests are released at different time instants, and no specified depot, as the problem is solved with arbitrary initial distributions of vehicles. On the whole, in Sect. 3.3, a meaningful practical benefit to the industrial process is documented.

## 2   MIP-Based Heuristic

In this section we recall the MIP-heuristic developed in [1]. The rectilinear topology allows us to represent the locations as a set of integers $I = \{i_1, \ldots, i_n\}$, corresponding to the location coordinates along the line. Now, given a subset of requests $P \subseteq R$, we define a minimal interval $[a(P), b(P)]$ that spans between the leftmost and rightmost pickup/delivery location involved in $P$; formally

$$a(P) = \min_{r \in P}\{p_r, d_r\} \qquad\qquad b(P) = \max_{r \in P}\{p_r, d_r\}$$

To illustrate the above operation, let us consider the following

*Example 1.* Let $I = \{1, 2, \ldots, 13, 14\}$ and $R$ as in Table 1

**Table 1.** Data for example 1.

| Request | $p_r$ | $d_r$ | $l_r$ |
|---------|-------|-------|-------|
| 1       | 5     | 2     | 1     |
| 2       | 1     | 3     | 2     |
| 3       | 13    | 6     | 2     |
| 4       | 5     | 14    | 3     |
| 5       | 4     | 3     | 1     |

Assuming $P = \{1, 4, 5\}$, stations $d_1 = 2, p_4 = 5, d_5 = 3$ (stations $p_1 = 5, d_4 = 14, p_5 = 4$) are the leftmost (rightmost) involved in those requests; hence, $a(P) = 2, b(P) = 14$.

We call any such interval a *span* of $P$. Spans are no more than the pairs of stations involved in $R$, namely $\leq \frac{q(q-1)}{2} = O(q^2)$ with $q = \min\{n, 2|R|\}$. Let $S$ indicate the set of all those spans, and denote by $w^s = b^s - a^s = b(P) - a(P)$ the width of the $s \in S$ corresponding to $P \subseteq R$. In our example $n = 14$ implies 91 possible spans in the worst case and the span $s = [2, 14]$ has $w^s = 14 - 2 = 12$.

Our two-stage heuristic for the R-VRPPD is based on the following decomposition principle:

1. Assign each request in $R$ to a span in a set $S^* \subseteq S$ with $|S^*| = |K| = m$;
2. Assign a vehicle $k \in K$ to each $s \in S^*$ and schedule its requests.

In stage 1, we cluster requests into spans of convenient width — the larger the span the fewer the requests — in order to balance the estimated operation time (travel + load/unload).

In stage 2, we determine elementary schedules where each vehicle "sweeps" from end to end the span it is assigned to, changing direction a minimum amount of times. Details on the two stages are given in next sections.

An important property of this decomposition is that a solution can easily be reconfigured at run time against possible vehicle capacity reduction, as well as urgent deliveries or other exceptions. This property will be fruitfully exploited when applying the heuristic in a dynamic environment (see Sect. 3.3).

### 2.1 Stage 1: Span Selection and Request Assignment

To simplify the method description, let us for the moment make a rough approximation (to be refined later) of the time a vehicle would take to cover a span $s$. We consider

 (i) the span width $w^s$: the larger the $w^s$, the longer will expectedly be the transfer time needed to cover its requests.
 (ii) the overall volume $l^s$ of requests assigned to $s$: large $l^s$ means long time spent in load/unload operations;

While $w^s$ only depends on $s$, the volume $l^s$ depends on the requests assigned to $s$. Let $R^s \subseteq R$ denote the set of requests whose pick-up and delivery stations fall within the ends of $s$. Then:

$$l^s = \sum_{r \in R^s} l_r x_r^s$$

where $x_r^s = 1$ if $r$ is assigned to $s$ and 0 otherwise. Using these variables, plus the activation variables $y^s \in \{0, 1\}$ with $y^s = 1$ if and only if $s \in S$ is chosen, we can select the $m$ spans via the following MIP (see also [1]):

$$\min_{z \in \mathbb{R}} \; z$$
$$\text{s.t.}$$
$$z - \frac{w^s}{v} y^s - 2u \sum_{r \in R^s} l_r x_r^s \geq 0 \quad s \in S$$
$$\sum_{s \in S_r} x_r^s = 1 \quad r \in R$$
$$x_r^s \leq y^s \quad s \in S, r \in R^s$$
$$\sum_{s \in S} y^s \leq m$$
$$\sum_{r \in R^s} l_r x_r^s \leq c \quad s \in S$$
$$x_r^s, y^s \in \{0, 1\} \quad s \in S, r \in R^s \tag{1}$$

where $S_r \subseteq S$ contains all the spans that can accommodate request $r$, and $u, v$ are the nominal load/unload time and vehicle speed, respectively. Coupled with the objective function, the first set of constraints defines $z$ as the largest among the completion time estimations of the requests attributed to a span. The second set of constraints enforces that every request is assigned to one span. The third selects a span when assigned a request. The remaining constraints ensure that the spans selected do not exceed available vehicles, and that vehicle capacity is in turn never exceeded.

The first four lines of program (1) have the aspect of an $m$-centre model, where spans corresponds to facilities, or centroids, and requests to demand points. In brief, an $m$-centre is a set $S^*$ of $m$ points (called centroids) that minimizes the largest distance between any demand point and the closest point in $S^*$: for a survey on $m$-centre problems see e.g. [4]. Our model differs from this problem in the distance function, that includes centroid costs (span lenghts) and, as said, for the additional knapsack constraints (vehicle capacities).

Once solved, program (1) returns:

- a set $S^*$ of $m$ spans;
- a partition $\{R^s : s \in S^*\}$ of the requests set to be assigned to vehicles in the subsequent stage 2;
- a lower bound $z^*$ to the minimum time necessary to schedule all the requests in $R$.

Let us now refine the initial estimation of transfer time, that we assumed proportional to the span width $w^s$: this estimate is in fact reasonable as far as the vehicle covers the span in one direction only, but becomes poor if it has to go back and forth. To obtain a better estimate, modify program (1) by distinguishing the requests in $R^s$ according to whether they require a transfer from left to right (forward requests, $F_t^s$) or from right to left (backward requests, $B^s$). Precisely

$$F^s = \{r \in R^s : p_r < d_r\} \qquad B^s = \{r \in R^s : p_r > d_r\}$$

We then add 0–1 variables $f^s, b^s$, subject to

$$\begin{aligned} x_r^s \leq f^s \leq y^s \qquad & r \in F^s \\ x_r^s \leq b^s \leq y^s \qquad & r \in B^s \end{aligned} \qquad (2)$$

for any $s \in S$. That is, span $s$ is used if a vehicle uses it either in forward or in backward direction (or both); conversely, if not used in forward (backward) direction, the span cannot be assigned any forward (backward) request. The constraint of model (1) that ensures to use $\leq m$ spans is maintained, but the first set of inequalities, defining the max completion time, is replaced by

$$z - \frac{w^s(f^s + b^s)}{v} - 2u \sum_{r \in R^s} l_r x_r^s \geq 0 \qquad (3)$$

for any candidate span $s \in S$. Thus, if the span is assigned requests in one direction only, the estimated travel time is proportional to $w^s$ as in model (1), otherwise is proportional to $2w^s$.

## 2.2   Stage 2: Vehicle Assignment and Scheduling

By our initial assumption on scheduling, any vehicle has to move end-to-end in the assigned span minimizing direction changes. This elementary idea can be implemented by different policies, for example:

– *Policy 1*: the vehicle picks up parts in the same order as it finds them in the travel, and delivers them in the same order as destinations are encountered.
– *Policy 2*: the vehicle first operates all the pick-ups in the order found in the travel; once all parts are collected, the vehicle delivers them in the order in which destinations are encountered.

Policy 1 also aims at minimizing the part stack in the vehicle, while the goal of Policy 2 is to minimize the work-in-process at the plant stations and the time each vehicle waits for another: in fact, with Policy 2, the pick-up operations precede every delivery, so we use from the very beginning the whole vehicle capacity available to free stations as far as possible. Tests conducted in [1] demonstrated that the best scheduling method follows Policy 1. So let us next detail the way it works.

For any $s \in S^*$, let $K^s$ contain the vehicles whose capacity is sufficient to accommodate all the pick-ups in $s$. Suppose then that span $s$ is assigned to vehicle $k \in K^s$, which starts from position $\xi_k$. To describe the scheduling algorithm implemented for $k$, we partition again $R^s$ into forward and backward requests $F^s, B^s$ as in Sect. 2.1. If $F^s$ is non-empty, let $p_F$ ($d_F$) denote the leftmost (rightmost) pick-up (delivery) location among its requests. Similarly, for non-empty $B^s$, let $p_B$ ($d_B$) denote the rightmost (leftmost) pick-up (delivery) location among its request. In example 1, one has $p_F = 1, d_F = 14, p_B = 13, d_B = 2$.

Assume both $F^s$ and $B^s$ non-empty. Then

– if $\xi_k < a^s$ (that is, the vehicle starting point lies to the left of $s$), $k$ will first fulfill forward requests, then backward ones;
– if $\xi_k > b^s$, $k$ will first fulfill backward requests, then forward ones.

If instead $a^s \leq \xi_k \leq b^s$, the schedule is constructed distinguishing three cases

i) $\xi_k < d_B$: the vehicle first fulfills the backward requests lying to its left, then changes direction and fulfills all the forward requests, then finally gets back to the remaining backward ones;
ii) $\xi_k > d_F$: the vehicle first fulfills the forward requests lying to its right, then changes direction and fulfills all the backward requests, then finally gets back to the remaining forward ones;
iii) $d_B \leq \xi_k \leq d_F$: we try both the above schedules and select for the vehicle the shortest one.

Finally, if $F^s = \emptyset$, then $k$ has only to fulfill backward requests, so it first moves to $p_B$ and then proceeds to fulfill all requests. A similar schedule with first pick-up in $p_F$ is constructed if $B^s = \emptyset$.

Once, via the rules described above, we have collected the scheduling times $\tau_k^s$ for each $s \in S^*$ and $k \in K^s$, we match vehicles and spans (and therefore vehicles and schedules) once for all, seeking to minimize the makespan $C_{\max}$ of all the requested services. This corresponds to solve a BOTTLENECK MATCHING PROBLEM [7], [3]: unlike ordinary matching, where the weight of a matching is the sum of the arc weights it consists of, a matching is here weighted by the largest of its arc weights. In our case, the problem is defined on a bipartite graph with node sets $K$, $S^*$, and arcs $(k, s)$ defined for every $s \in S^*$, $k \in K^s$ and weighted by $\tau_k^s$. An optimal solution to this problem can be found in $O(m^3)$ time [10].

## 3  Industrial Case Study

LFoundry (http://www.lfoundry.com) is a company that manufactures high-tech micro-electronic components and devices in a plant located in Avezzano, Italy. The heart of the plant is the *clean room*, equipped with some 700 machines clustered according to group technology with a layout as in the scheme of Fig. 1. Basically, a central aisle distributes production flows between a warehouse and a machine area, and within machine sub-areas. Machines are accessed via fourteen dedicated aisles, that lay orthogonally to the main one and are provided at their heads with a rack for in-process inventory. Parts are moved from rack to rack by a fleet of manually operated carts that commute back and forth along the main aisle, and by individual operators between the machines and the relevant head racks.

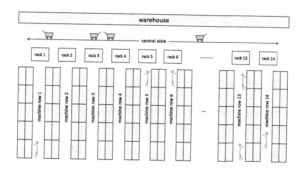

**Fig. 1.** Scheme of the clean room of LFoundry [1].

Production is organized in small lots of identical size. Depending on product type, a lot must undergo a very large number of operations (roughly from 250 to 750). Machines can be tooled to perform specific operations, and each lot typically has the same operation type repeated many times. This, broadly speaking, gives the manufacturing system the aspect of a (very big and complex) flexible job-shop and controlling work-in-process (WIP) is one crucial issue to optimize production. Lots not currently worked by some machine, wait for operation in the racks positioned in the main aisle. Radio frequency identification devices (RFID) continuously feed the information system with the actual position of every lot waiting for or worked by a machine. These conditions suggest that a proper vehicle routing/scheduling could improve system performance and reduce the WIP.

### 3.1  Test-Bed and Computational Settings

LFoundry provided us with very large and detailed data samples on clean room operation. Each sample describes a work shift of over 6 h, and is formed by a series of snapshots (one per minute) of request status. Besides other information, a snapshot gives the requests pending, those in process and newly arrived ones, all recorded according to the actual practice of cart routing and scheduling. On average, 55.7% of the requests are short-distance, that is, cover < 6 stations (i.e., racks) in the aisle, and only 1.8% are long-distance, that is, range over all the 14 stations. More features are given in Table 2 for the five samples used in the numerical experience.

Table 2. Sample work shift features.

| Sample | # Snapshots | Requests per snapshot | | | | Requests per sample |
|--------|-------------|------|------|------|------|---------------------|
|        |             | Min | Max | Avg | $\sigma$ |                |
| 1 | 374 | 38 | 96 | 91 | 0,2 | 3.110 |
| 2 | 351 | 32 | 95 | 86 | 0,6 | 3.076 |
| 3 | 343 | 41 | 96 | 80 | 0,7 | 2.984 |
| 4 | 342 | 45 | 94 | 86 | 0,5 | 2.902 |
| 5 | 347 | 54 | 95 | 82 | 0,8 | 2.943 |
| Total | 1.757 | | | | | 15.015 |

Any two adjacent stations in the clean room are separated by 10 m. A fleet of eight manually operated vehicles (carts) is used for transportation, each with a loading capacity of 12 lots. The (constant) cart speed is 1.2 m/s and per-lot load/unload time (15 s).

The discussion of the experimental campaign is divided into two sections:

i) In Sect. 3.2 we show the comparison of our two-stage MIP heuristic with Clarke and Wright's heuristic (CW) on a test-bed of 50 instances. Each instance is given by a randomly selected snapshot, has between 30 and 41 requests (33.6 on average), and counts among 56 and 82 lots (70.7 on average). According to the input requirement of CW, we assume that all requests are known a priori and that carts start/end their routes at a (fictitious) depot located in a position $h$ between the third and the fourth station.

ii) In Sect. 3.3 we present the results achieved by our approach on the five samples described in Table 2, and compare them to current practice. To be compliant with the recorded dynamic evolution of the process, in these experiments we assume requests arrivals over time according to available data.

Algorithms were encoded in Python 3.7.6. and run on an Intel core 7, 1.8–1.99 GHz processor, 16 GB RAM, operation system Windows 10 Pro 64-bit version 2004. The MIP solver adopted is Gurobi 9.0.2 with default settings using 8 threads.

## 3.2   Comparison with a Clarke and Wright Heuristic

As a benchmark for assessing the MIP-based heuristic performance, we implemented a standard *cluster-first route-second* algorithm based on the classical Clarke and Wright framework [5] for the VRP, integrated with a *2-opt* local search for the TSP [6]. This is a two-phase method in which the former step is devoted to cluster requests according to *savings* $\sigma$ and assign clusters to carts, while the latter phase tries to minimize the route length of each cart by performing simple swaps among route components. The method pseudo-code is given in Algorithm 1 and explained hereafter.

---

**Algorithm 1.** Clarke and Wright heuristic.

---

```
 1: procedure CW(I, K, R)
 2:     C = ∅
 3:     C ← init(R)
 4:     repeat
 5:         σ_max ← −∞
 6:         for C_i ≠ C_j ∈ C : L_i + L_j ≤ c do
 7:             (d̄, p̄) ← getPD(C_i, C_j)
 8:             σ ← w_{d̄h} + w_{hp̄} − w_{d̄p̄}
 9:             if σ_max < σ then
10:                 σ_max ← σ, (ι, κ) ← (i, j)
11:         C ← update(C, ι, κ)
12:     until |C| ≤ m
13:     for C_i ∈ C do
14:         iter ← 0
15:         repeat
16:             (C̃_i, iter) ← rand(C_i, iter)
17:             repeat
18:                 ω_max ← −∞, flag ← 0
19:                 for e, ē ∈ Ẽ_i : e ∩ ē = ∅ do
20:                     ω ← gain(C̃_i, e, ē)
21:                     if ω_max < ω ∧ feas(C̃_i, e, ē) then
22:                         ω_max ← ω, (μ, ν) ← (e, ē)
23:                 (C̃_i, flag) ← swap(C̃_i, μ, ν)
24:             until ¬ flag
25:             if getCT(C̃_i) < getCT(C_i) then
26:                 C_i ← C̃_i
27:         until iter > N
28:     z ← max_{C_i∈C}{getCT(C_i)}
```

---

First, the set $C$ is populated with an initial cluster $C_r = \{h, p_r, d_r, h\}$ per $r \in R$. Clusters are treated as set of positions, but a cycle $(h, p_r, d_r, h)$ can be easily derived starting from the depot in position $h$, passing through $p_r$ and $d_r$, then returning to the depot. Moreover, $L_r$ is initialized to $l_r$.

For each pair $C_i$ and $C_j$ of different clusters which involve a lot volume fitting the capacity of cart $c$, we compute the saving $\sigma$ on the basis of the distance $w_{\alpha\beta} = |\alpha - \beta|$

that separates positions $\alpha$ and $\beta$ in a cluster. Let $\bar{d} \in C_i$ and $\bar{p} \in C_j$ respectively be the delivery and pickup locations with minimum distances $w_{\bar{d}h}$ and $w_{h\bar{p}}$ from the depot $h$. Then, $\sigma = w_{\bar{d}h} + w_{h\bar{p}} - w_{\bar{d}\bar{p}}$ measures the largest potential saving one achieves by merging the cycles derived from $C_i$ and $C_j$, that is, by moving from $\bar{d}$ directly into $\bar{p}$ without returning to (and departing again from) depot $h$. Among all such merging operation, we implement the one identified by the clusters indices $\iota$ and $\kappa$ that achieve the largest saving $\sigma_{\max}$: $C$ is then properly updated by appending pickup/delivery locations of $C_\kappa$ in $C_\iota$ and by deleting $C_\kappa$. Finally, $L_\iota$ is increased by $L_\kappa$.

Typically, this clustering phase is iterated while $\sigma_{\max} \leq 0$ is found. However, as a limited number of carts is available and considering the goal of makespan minimization, we terminate this phase when no more than $m$ clusters are gathered, trying in this way to balance the cart loads and help reduce the makespan. Notice that, in this experimental setting, assigning clusters to carts is trivial because carts are identical and share the starting position $\xi_k = h$; by the way, observe also that this assumption allows us to skip the bottleneck matching step in the MIP heuristic.

Once clusters are defined, the algorithm proceeds with the routing phase. For each cluster $C_i$ obtained, we repeat building a cycle $\tilde{C}_i = (h, \ldots, h)$ up to $N$ times by function $rand$, which randomly sorts pickup and delivery locations in $C_i$. A cycle $\tilde{C}_i$ is accepted if respects pickup and delivery precedences. We define $\tilde{E}_i$ as the set of ordered pairs of adjacent locations $e = (\alpha, \beta)$ in $\tilde{C}_i$, and $\sum_{e \in \tilde{E}_i} w_e$ as the cycle length. For any two pairs $e = (\alpha, \beta)$ and $\bar{e} = (\bar{\alpha}, \bar{\beta})$ not sharing a location, function $gain$ computes the cycle length reduction $\omega$ obtained by swapping locations $\beta$ and $\bar{\alpha}$ in the cycle. Note that this swap reverses the path $(\beta, \ldots, \bar{\alpha})$ into $(\bar{\alpha}, \ldots, \beta)$ and brings from cycle $(h, \ldots, \alpha, \beta, \ldots, \bar{\alpha}, \bar{\beta}, \ldots, h)$ to $(h, \ldots, \alpha, \bar{\alpha} \ldots, \beta, \bar{\beta}, \ldots, h)$. Thus, precedence conflicts must be checked to ensure the new cycle feasibility. The feasible swap that defines $\omega_{\max}$ is then applied to $\tilde{C}_i$, and new swaps are further tried until no improvement is found. The cycle underlying $C_i$ is updated if its completion time is larger than the completion time of cycle $\tilde{C}_i$. Finally, the makespan $z$ is taken as the largest completion time among all $C_i$.

Table 3 illustrates a comparison between the two-stage MIP heuristic and the CW heuristic with $N = 500$. For each of the 50 instances we report:

- number of requests $|R|$ and total load $\sum_{r \in R} l_r$;
- gap between the best known objective and the current lower bound at termination reported by Gurobi in stage 1;
- the makespan $C_{max}$ achieved and the total CPU time in seconds, for both heuristics.

For the MIP heuristic the CPU time substantially coincides with Gurobi elapsed time, as the time required by stage 2 is always negligible. The last column reports the percentage relative gap $\frac{C_{max}(CW) - C_{max}(MIPH)}{C_{max}(MIPH)}$. In this experiment the time limit for Gurobi is set to 120 seconds (when it is reached, the corresponding entry reads 'tlim'): this time limit was suggested by the time allowed to make routing decision during operation.

The statistics obtained clearly show that the MIP heuristic sensibly outperforms the CW heuristic in terms of solution quality. In fact the proposed method, tailored to the rectilinear topology, always improves the makespan computed with CW by a meaningful amount, larger than 15% in 37 out of 50 cases and larger than 20% in 23 cases. On average, the improvement obtained is 19.60%.

**Table 3.** MIP heuristic vs. CW heuristic

| $|R|$ | Total load | MIP heuristic | | | Clarke-wright | | MIP vs CW |
|---|---|---|---|---|---|---|---|
| | | Gurobi opt. gap | $C_{max}$ | Time | $C_{max}$ | Time | % gap |
| 30 | 56 | 0 | 295 | 6.84 | 396.67 | 0.39 | 34.46 |
| 31 | 58 | 0 | 310 | 6.94 | 396.67 | 0.38 | 27.96 |
| 32 | 71 | 0 | 340 | 42.73 | 415 | 0.42 | 22.06 |
| 33 | 61 | 0 | 325 | 12.11 | 380 | 0.47 | 16.92 |
| 32 | 67 | 3.97 | 325 | tlim | 380 | 0.44 | 16.92 |
| 31 | 65 | 1.65 | 325 | tlim | 380 | 0.42 | 16.92 |
| 30 | 62 | 2.48 | 310 | tlim | 396.67 | 0.38 | 27.96 |
| 33 | 74 | 0 | 325 | 86.91 | 406.67 | 0.47 | 25.13 |
| 32 | 68 | 3.12 | 325 | tlim | 421.67 | 0.39 | 29.74 |
| 34 | 69 | 0.78 | 340 | tlim | 390 | 0.55 | 14.71 |
| 34 | 71 | 0 | 340 | 74.92 | 390 | 0.44 | 14.71 |
| 32 | 73 | 3.05 | 325 | tlim | 391.67 | 0.56 | 20.51 |
| 33 | 74 | 1.53 | 325 | tlim | 380 | 0.48 | 16.92 |
| 33 | 73 | 5.22 | 340 | tlim | 380 | 0.61 | 11.76 |
| 33 | 74 | 7.41 | 340 | tlim | 380 | 0.48 | 11.76 |
| 34 | 72 | 0 | 325 | 66.19 | 393.33 | 0.42 | 21.03 |
| 31 | 73 | 1.5 | 325 | tlim | 373.33 | 0.41 | 14.87 |
| 32 | 73 | 0 | 325 | 75.09 | 380 | 0.53 | 16.92 |
| 36 | 69 | 4.72 | 325 | tlim | 373.33 | 0.52 | 14.87 |
| 35 | 66 | 3.97 | 325 | tlim | 405 | 0.5 | 24.62 |
| 34 | 64 | 0 | 325 | 5.08 | 408.33 | 0.48 | 25.64 |
| 36 | 63 | 4 | 308.33 | tlim | 408.33 | 0.5 | 32.43 |
| 41 | 71 | 4.44 | 340 | tlim | 406.67 | 0.64 | 19.61 |
| 39 | 71 | 1.49 | 340 | tlim | 438.33 | 0.61 | 28.92 |
| 38 | 72 | 5.84 | 336.67 | tlim | 406.67 | 0.59 | 20.79 |
| 39 | 80 | 4.86 | 355 | tlim | 438.33 | 0.59 | 23.47 |
| 38 | 82 | 6.16 | 355 | tlim | 423.33 | 0.63 | 19.25 |
| 36 | 80 | 4.93 | 340 | tlim | 421.67 | 0.47 | 24.02 |
| 34 | 75 | 0 | 340 | 64.56 | 413.33 | 0.48 | 21.57 |
| 36 | 77 | 0 | 340 | 51.17 | 398.33 | 0.44 | 17.16 |
| 35 | 74 | 0 | 340 | 64.95 | 381.67 | 0.47 | 12.25 |
| 31 | 70 | 0 | 325 | 88.14 | 381.67 | 0.41 | 17.44 |
| 31 | 69 | 0 | 325 | 75.27 | 396.67 | 0.41 | 22.05 |
| 30 | 64 | 0 | 310 | 11.69 | 396.67 | 0.34 | 27.96 |
| 30 | 65 | 0 | 325 | 42.7 | 381.67 | 0.45 | 17.44 |
| 32 | 66 | 1.65 | 310 | tlim | 405 | 0.45 | 30.65 |
| 33 | 67 | 5.65 | 325 | tlim | 381.67 | 0.41 | 17.44 |
| 33 | 69 | 7.75 | 340 | tlim | 405 | 0.45 | 19.12 |
| 32 | 64 | 2.44 | 310 | tlim | 376.67 | 0.39 | 21.51 |
| 33 | 68 | 7.81 | 325 | tlim. | 396.67 | 0.47 | 22.05 |
| 33 | 71 | 4.69 | 321.67 | tlim | 393.33 | 0.33 | 22.28 |
| 33 | 70 | 2.34 | 321.67 | tlim | 408.33 | 0.42 | 26.94 |
| 33 | 68 | 0 | 325 | 71.38 | 378.33 | 0.45 | 16.41 |
| 33 | 75 | 2.96 | 340 | tlim | 365 | 0.41 | 7.35 |
| 33 | 76 | 0 | 340 | 71.25 | 380 | 0.36 | 11.76 |
| 33 | 78 | 0 | 336.67 | 89.88 | 380 | 0.41 | 12.87 |
| 34 | 81 | 2.16 | 355 | tlim | 365 | 0.41 | 2.82 |
| 35 | 80 | 0 | 336.67 | 72.66 | 396.67 | 0.47 | 17.82 |
| 36 | 78 | 4.96 | 340 | tlim | 375 | 0.45 | 10.29 |
| 37 | 79 | 4.96 | 340 | tlim | 375 | 0.48 | 10.29 |

Notice also that such a relevant improvement is observed even if Gurobi reached the time limit in 30 out of 50 cases. Indeed, in a preliminary experiment we investigated the sensitivity of the MIP heuristic with respect to the quality of the solution computed in stage 1. In particular, we have run the experiment with a large time limit so as to let Gurobi certify optimality (this may take up to 1 hour) and we experienced that this does not yield a significant improvement to the final $C_{max}$. In other words, a good performance of the two-stage heuristic is already achieved with near optimal solutions of stage 1. These turned out to be quickly computed by Gurobi primal heuristic (with default settings) at very early stages of the branch-and-cut algorithm.

Such a well-behaviour of the MIP-based heuristic supported the investigation of the potential benefits of its implementation in the LFoundry production process. This is illustrated in the next section.

### 3.3   Practical Impact

This subsection follows again [1]. It is divided in three parts: *Current practice*, *Dynamic implementation* and *Results*, and describes experiments whose aim is to illustrate the behaviour of the MIP heuristic proposed, when applied to a very large number of requests arriving over time. As observed, both to be time-compliant and to leave enough freedom to manage exceptions that may possibly occur, route operation imposes a repeated application of the method to subsets of requests of adequately small size. This implies

- the addition of release dates to the R-VRPPD, that is, time windows within which requests can be considered available for processing;
- the absence of a central cart depot, as the algorithm must now input initial cart positions anywhere in the line.

The latter issue is approached by a particular BOTTLENECK MATCHING PROBLEM, as explained at the end of Sect. 2.2. The former issue requires to deal with details not particularly demanding in terms of computation time, but still needing consideration in order to correctly concatenate subsequent problem solutions. All these details are described in the *Dynamic implementation* sub-section.

**The Current Practice.** The current policy is based on a fixed cart routing, assessed via statistics identifying the most frequent transportation requests. Specifically, three static routes are defined: the longest, covering all the fourteen stations in the main aisle; the shortest, covering its central part; and an intermediate one, strictly contained in the former and strictly containing the latter. The cart fleet is then three-partitioned and statically assigned to those routes.

**Dynamic Implementation.** We have implemented our two-stage heuristic in a dynamic environment so as to evaluate its impact on the LFoundry production process. Now we have a discrete time horizon $\mathcal{T} = 1, \ldots, T$ and denote by $R_t$ the set of requests that, at some time $t \in \mathcal{T}$, (i) are not yet picked up and (ii) have lots available at the respective stations. Cart $k$ is in general available from some time $t_k \geq t$ on, as it is supposed to conclude the deliveries assigned in the previous time lap. At that time, the cart will have a particular position $\xi_k$ in the aisle and a particular residual capacity $\bar{c}_k$. As previously noticed, this means that cart initial positions generally differ from one another, so making the one-depot assumption inapplicable and requiring, on the other hand, to solve in stage 2 the BOTTLENECK MATCHING problem described at the end Sect. 2.2.

The experience was carried out with the following procedure, see also the scheme in Fig. 2:

1. *Initialize.* Construct the list $L$ of all the requests in a sample, each with the time of its arrival. Let $t = 0$ and let $R_0$ contain the requests of the first snapshot.
2. *Cycle.* Run the algorithm of Sect. 2 on the requests of $R_t$, let $C_{max}$ and $C_{min}$ be the time length of the longest and the shortest route of a cart. Take note of the status (positions $\xi_k$ and minimum ready time $t_k$) of all the carts at $t + C_{max}$.
3. *Update and Repeat.* Set $t := t + C_{min}$, create $R_t$ including in it all the requests of $L$ that arrived within $[t - C_{min}, t]$. Taking into account cart status, go back to step 2 and repeat until all the requests in $L$ have been fulfilled.

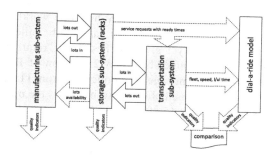

**Fig. 2.** Scheme of computational experience [1]: plant sub-systems (grey blocks), software components (white blocks), physical flows (solid arrows), main information flows (dashed arrows).

**Results.** We compare the makespan determined for the requests in $L$ with that from the current practice. The number of variables in the MIP (1)-(3) ranges from a minimum of 2.667 to a maximum of 6.767, that of constraints from 2.669 to 6.333. The final outcome is reported in Table 4. All in all, we processed 15.015 requests for a total work time of 1.757 min, see column 2. The result of a numerical test based on model (1) is reported in columns 3–4, another test with the model updated by (2), (3) gives the values in columns 5–6. The former (the latter) test results in a rough 9% (30%) improvement over actual practice, amounting in absolute terms to over two hours and $\frac{1}{2}$ (eight hours and $\frac{3}{4}$) potentially saved in five work shifts.

**Table 4.** Algorithm performance. Minutes to complete operation, actual duration vs. tests with model (1) and update (2, 3), and percentage reductions.

| Sample id. | Actual duration | Improvement (test) | | | |
|---|---|---|---|---|---|
| | | Model (1) | | Update (2, 3) | |
| 1 | 374 | 344 | 8,1% | 252 | 32,7% |
| 2 | 351 | 318 | 9,3% | 254 | 27,5% |
| 3 | 343 | 308 | 10,3% | 248 | 27,7% |
| 4 | 342 | 318 | 7,0% | 246 | 28,1% |
| 5 | 347 | 312 | 10,2% | 231 | 33,5% |
| Total | 1.757 | 1.599 | 9,0% | 1.231 | 29,9% |

The second test gives definitely better results at a price, however, of larger CPU times. In fact, every cycle (step 2 of the above procedure) took in this case about 46,4 s CPU time on average, ranging from a minimum of 4,9 to a maximum of 147,8 s. CPU times of the first test were quite shorter: 3,6 s on average with a minimum of 1,4 s and a maximum of 6,9 s.

We also tested our optimization method under Policy 2 of Sect. 2.2. Over all samples, this scheduling policy returned makespan of 1.477 min, that is 246 min (about 20%) worse than Policy 1 but still 280 min (16%) better than actual operation.

Besides makespan, it is interesting to observe how cart workloads are distributed with the proposed approaches. Table 5 gives, per sample, the mileage covered by the carts. Also under this respect Policy 1 gives the best result, returning an average cart mileage 15,7% shorter than that attained by Policy 2 (last row of Table 5). As for route length distribution, the largest cart mileage with Policy 1 is from 11,6 to 12,9 km shorter than with Policy 2. In absolute terms, Policy 1 also improves the mileage unbalance of Policy 2 of amounts ranging between 15,4% and 16,8%. However, the relative unbalance

$$U_R = \frac{\text{longest route} - \text{shortest route}}{\text{longest route}}$$

is quite large with both policies, from 48.0% to 51.6% (51.2% with Policy 2). To correctly read this information one has however to consider that program (1)-(3) tends to assign less lots to longer routes, so as to compensate travel and load/unload time.

**Table 5.** Mileage balance. Kilometres per cart to complete operation: average, maximum and unbalance (diff.) per sample and scheduling policy.

| Sample id. | Cart mileage (km) | | | | | |
|---|---|---|---|---|---|---|
| | Policy 1 | | | Policy 2 | | |
| | Avg. | Max | Diff. | Avg. | Max | Diff. |
| 1 | 41.4 | 63.9 | 31.5 | 49,5 | 76.7 | 37.8 |
| 2 | 42.6 | 64.2 | 30.8 | 50.6 | 77.1 | 37.0 |
| 3 | 39.3 | 62.8 | 31.7 | 46.2 | 75.4 | 38.0 |
| 4 | 39.6 | 62.1 | 31.2 | 46.5 | 74.5 | 37.4 |
| 5 | 38.5 | 63.8 | 31.5 | 48.9 | 76.6 | 37.5 |
| Tot. avg. | 41.2 | | | 48.9 | | |

Finally, a simple test of the robustness of the method against parameter variation was done with a new run of all samples, randomly choosing cart speed and load/unload time uniformly picked in $1, 2 \pm 0, 1$ m/s and $15 \pm 2$ s. With both Policy 1 and 2, this experiment returned a makespan substantially close to the values found with constant parameters.

# 4 Conclusions

We have presented a cluster-first-route-second decomposition framework for a Vehicle Routing Problem with Pickups and Deliveries on a rectilinear layout. Unlike classical graph-based approaches, the clustering stage is carried out by solving a $m$-CENTRE-type MIP model; then, vehicles arbitrarily deployed along the line are assigned with routes by solving a BOTTLENECK MATCHING problem. An extensive computational experience has shown that this heuristic outperforms a tailored implementation of the classical Clarke and Wright heuristic. Moreover, even primal feasible solutions of the clustering MIP, quickly computed by the off-the-shelf solver, turned out to yield high quality final solutions. As such, our method looks promising for industrial implementation, which is currently under assessment by LFoundry production managers.

# References

1. Arbib, C., Ranjbar, F.K., Smriglio S.: A dynamic dial-a-ride model for optimal vehicle routing in a wafer fab. In: Proceedings of the 10th International Conference on Operations Research and Enterprise System, Online Streaming, 4–6 February 2021, pp. 281–286 (2021)
2. Berbeglia, G., Cordeau, J.F., Gribkovskaia, I., Laporte, G.: Static pickup and delivery problems: a classification scheme and survey. TOP **16**, 1–31 (2007). https://doi.org/10.1007/s11750-007-0009-0
3. Burkard, R., Derigs, U.: The bottleneck matching problem. In: Assignment and Matching Problems: Solution Methods with FORTRAN-Programs. Lecture Notes in Economics and Mathematical Systems, vol. 184, pp. 60–71. Springer, Berlin-Heidelberg (1980). https://doi.org/10.1007/978-3-642-51576-7_5
4. Calik, H., Labbé, M., Yaman, H.: $p$-center problems. In: Laporte, G., Nickel, S., da Gama, F.S. (eds.) Location Science, pp. 79–92. Springer, Cham (2015). https://doi.org/10.1007/978-3-319-13111-5_4
5. Clarke, G., Wright, J.W.: Scheduling of vehicles from a central depot to a number of delivery points. Oper. Res. **12**, 568–581 (1964)
6. Croes, G.A.: A method for solving traveling salesman problems. Oper. Res. **6**, 791–812 (1958)
7. Garfinkel, R.: An improved algorithm for the bottleneck assignment problem. Oper. Res. **19**(7), 1747–1751 (1971)
8. Guan, D.J.: Routing a vehicle of capacity greater than one. Discrete Appl. Math. **81**, 41–57 (1998)
9. Ho, S., Szeto, W., Huo, Y., Leung, J., Petering, M., Tou, T.: A survey of dial-a-ride problems: literature review and recent developments. Transp. Res. B **111**, 395–421 (2018)
10. Pundir, P., Porwal, S., Singh, B.: A new algorithm for solving linear bottleneck assignment problem. J. Inst. Sci. Technol. **20**(2), 101–102 (2015)

11. Ralphs, T., Kopman, L., Pulleyblank, W., Trotter, L.E.: On the capacitated vehicle routing problem. Math. Program. Ser. B **94**, 343–359 (2003). https://doi.org/10.1007/s10107-002-0323-0
12. Vidal, T., Laporte, G., Matl, P.: A concise guide to existing and emerging vehicle routing problem variants. Eur. J. Oper. Res. **286**(2), 401–416 (2020)
13. Wang, F., Lim, A., Xu, Z.: The one-commodity pickup and delivery travelling salesman problem on a path or a tree. Networks **48**(1), 24–35 (2006)

# Author Index

Printed in the United States
by Baker & Taylor Publisher Services